THE GENIUS OF C. WARREN THORNTHWAITE, CLIMATOLOGIST-GEOGRAPHER

C. Warren Thornthwaite, about 1952.

The Genius of
C. Warren Thornthwaite,
Climatologist-Geographer

By John R. Mather
and Marie Sanderson

UNIVERSITY OF OKLAHOMA PRESS : NORMAN AND LONDON

BY JOHN R. MATHER

Climatology: Fundamentals and Applications (New York, 1974)
The Climatic Water Budget in Environmental Analysis (Lexington, Mass., 1978)
Water Resources: Distribution, Use, and Management (New York, 1984)
Global Change: Geographical Approaches (Tucson, 1991)
The Genius of C. Warren Thornthwaite, Climatologist-Geographer (Norman, 1995)

BY MARIE SANDERSON

Griffith Taylor: Antarctic Scientist and Pioneer Geographer (Ottawa, Ont., 1988)
Letters from a Soldier (with R. M. Sanderson) (Waterloo, Ont., 1993)
Prevailing Trade Winds: Weather and Climate in Hawaii (Honolulu, 1993)
The Genius of C. Warren Thornthwaite, Climatologist-Geographer (Norman, 1995)

Library of Congress Cataloging-in-Publication Data

Mather, John Russell, 1923–
  The genius of C. Warren Thornthwaite, climatologist-geographer / by John R. Mather and Marie Sanderson.
    p. cm.
  Includes bibliographical references and index.
  ISBN 0-8061-2787-2 (cloth : alk. paper)
  1. Thornthwaite, C. W. (Charles Warren), 1899–1963.
  2. Climatologists—United States—Biography. 3. Geographers—United States—Biography. I. Sanderson, Marie. II. Title.
  QC858.T48M38 1995
  551.6'092—dc20
  [B]   95-14239
                                                                     CIP

The paper in this book meets the guidelines for permanence and durability of the Committee on Production Guidelines for Book Longevity of the Council on Library Resources, Inc. ∞

Copyright © 1996 by the University of Oklahoma Press, Norman, Publishing Division of the University. All rights reserved. Manufactured in the U.S.A.

1 2 3 4 5 6 7 8 9 10

Contents

List of Illustrations		vii
Preface		ix
Acknowledgments		xiii
1	Early Years: 1899–1926	3
2	The University of Oklahoma: 1927–1934	17
3	Washington and the Soil Conservation Service: 1934–1946	34
4	The Move to Seabrook and an "Institute for Climatic Research"	72
5	Thornthwaite and Seabrook Farms: 1946–1954	83
6	C. W. Thornthwaite Associates	125
7	Thornthwaite and Academic Geography	140
8	International Activities: 1947–1958	155
9	The Laboratory and the Associates during the 1950s	167
10	The Last Years	182
11	Thornthwaite's Legacy	190
Thornthwaite's Published Works		203
Bibliography		213
Index		217

Illustrations

C. Warren Thornthwaite, about 1952	*Frontispiece*
Thornthwaite, at age two	6
Thornthwaite with his family, 1906	7
Thornthwaite and Denzil Slentz, 1923	9
Jack Seabrook, Thornthwaite, Maurice Halstead, and Russ Mather, 1953	81
Thornthwaite and daughter Elizabeth, 1953	86
Thornthwaite and early Cropmeter, 1952	91
Vegetative development of the garden pea	94
Stages of development of peas between nodes	95
Evaluating the development of corn, about 1952	96
Solar radiation, plant development, the development index, and potential evapotranspiration	98
Adjusting thermocouples to obtain air temperature profiles, 1953	110
Firing a smoke puffer to determine air turbulence	111
Pumping station 3 for wastewater disposal, Seabrook, N.J., 1951	117
Thornthwaite and Jack Seabrook discuss wastewater disposal, 1951	118
Depth of water applied to wastewater disposal area, Seabrook Farms, 1950	121
Anders Angström and Thornthwaite, 1955	129
Distribution of rated geographers in American universities and federal employment	153

Thornthwaite presiding over the Commission
   for Climatology of the World Meteorological
   Organization, 1957   164

Warren and Denzil Thornthwaite en route to
   Australia, 1957   165

Rudolf Geiger lecturing at the Laboratory
   of Climatology, Seabrook, N.J., 1950   169

Thornthwaite, Rudolf Geiger, and Heinz Lettau,
   1950   170

Thornthwaite, A. Austin Miller, and Takeshi
   Sekiguti, 1952   173

Thornthwaite receiving the Cullum Medal of the
   American Geographical Society from President
   Walter Wood   180

Thornthwaite family, Christmas, 1961   186

Preface

Charles Warren Thornthwaite possessed that rarely encountered spark of genius. One could argue that he was the world's most outstanding climatologist; in any case, he was unquestionably the most outstanding American climatologist of the twentieth century. He profoundly influenced the development of the modern field of climatology.

Thornthwaite became famous in geographic and climatic circles as the originator of a new, scientific classification of climate in the mid-twentieth century. A more permanent contribution to world research was his concept of potential evapotranspiration and his water balance model, which formed the basis of his climatic classification. The water balance model, which compares water need or potential evapotranspiration with water supply or precipitation, was first published by Thornthwaite in 1948 and has continued to be used by scientists in many disciplines for almost fifty years. Its simplicity and ease of verification have made the model a useful tool for hydrologists, forest and agricultural scientists, and engineers. The model has been used to estimate surface runoff in areas for which no runoff data exist and to extend the runoff record back in time. It provides data on soil moisture and evapotranspiration that are very useful for agriculture and forest scientists. The provision of quantitative estimates of water deficiency has similarly proved of practical use to crop scientists and irrigation engineers.

Perhaps Thornthwaite himself did not visualize the many future uses of his water budget technique. As a college professor, his original goal was to develop a new climatic classification; later, as a consulting climatologist, he began to realize that his novel concept of potential evapotranspiration and his water budget model had practical uses in many fields of science. He could thus be called the father of applied climatology. Although he had no formal training in instrument manufacture, Thornthwaite also earned recognition for the design of inexpensive, practical meteorological instruments. Many of his radiometers and wind sensors are still in use.

This book is an attempt to describe Thornthwaite's life and works and the impact of his writings on American and world science. It is written by two geographers who were his students and his friends. Marie Lustig (Sanderson) was a student at the University of Maryland in 1945–46, having come from the University of Toronto in Canada to work with O. E. Baker, who was then chairman of the Department of Geography at the University of Maryland but was also much involved with the Department of Agriculture in Washington, D.C. At that time, Thornthwaite lived in College Park, near the university, and worked in the Soil Conservation Service of the Department of Agriculture. He was asked by Baker to give an evening course in climatology, and Marie became one of his students. Thornthwaite was then working on "An Approach toward a Rational Classification of Climate" (published in 1948), and his ideas were exciting and innovative. Under his influence, Marie chose to do her master's thesis in climatology, and although she returned to Canada after graduation to pursue a career in climatology, the friendship between the two lasted until Thornthwaite's death in 1963.

John R. "Russ" Mather first met Warren Thornthwaite in 1948 when he went to Seabrook, N.J., to be interviewed for a job as a researcher at the newly established Laboratory of Climatology. He accepted the position, and he, too, fell under the spell of Thornthwaite's magnetism. During that first summer, he lived on the third floor of the Thornthwaite residence and shared Thornthwaite's long hours of work for six days a week. For Russ, the excitement of new challenges, research opportunities, and contacts with international climatologists who visited the laboratory made Seabrook impossible to leave. Russ obtained his doctorate from Johns Hopkins University while working at the laboratory with Thornthwaite and continued on the research staff of the organization for the sixteen remaining years of Thornthwaite's life. After Thornthwaite's untimely death, Russ became president of C. W. Thornthwaite Associates Laboratory of Climatology and managed that organization for nine more years.

The information in the book comes from a variety of sources: Thornthwaite's published works; his correspondence in the files of the Laboratory of Climatology and C. W. Thornthwaite Associates, Centerton, N.J.; recollections by family members and friends; and the authors' personal remembrances. We have made liberal use of quotes from Thornthwaite's letters and writings. In this way we hope to make more understandable the man's character as well as his scientific interests. We also quote from letters written to Thornthwaite by friends and colleagues, of whom many were well-known scientists like Carl Sauer, Isaiah Bowman, and Wladimir Köppen. In addition, the present-day recollections of Thornthwaite's friends and former associates, such as Ken Hare and C. C. Wallén, aid in evaluating the scientific and human legacy of the man.

We hope that this account of the life and work of Warren Thornthwaite will help to explain to workers in the fields of climatology and water resources research the many contributions of this famous climatologist and geographer. We hope that the book will serve also as an expression of our deep gratitude to the man who so profoundly influenced our own lives.

JOHN R. MATHER

Newark, Delaware

MARIE SANDERSON

Waterloo, Ontario

Acknowledgments

We would like to express our thanks to Fred Thornthwaite (brother), Elizabeth Higgins-Hallway (daughter), and Beth Haden (friend of the family) for information on Thornthwaite's life and to the Development Office of Central Michigan University for the material on pioneer days in Michigan written by Floyd Slentz, Thornthwaite's brother-in-law. We also wish to thank Fred Kniffen, Louisiana State University; George Carter, The Johns Hopkins University; Arnold Court, California State University, Northridge; Leslie Hewes, University of Nebraska; David Miller, University of Wisconsin–Milwaukee; Ken Hare, Toronto, Canada; Ben Garnier, Montreal, Canada; L. A. Ramdas, New Delhi, India; M. I. Budyko, Saint Petersburg, Russia; and C. C. Wallén, Stockholm, Sweden, for their personal recollections of Thornthwaite as an academician and scientist.

A copy of Thornthwaite's doctoral dissertation and information on his Berkeley days, and material from the Carl Sauer archives were kindly supplied by David Stoddard, University of California at Berkeley. Material on Wladimir Köppen and Rudolf Geiger was given to us by Rudolf Geiger, Munich, Germany, before his death. We are also grateful to the National Archives in Washington for making available the files of the early years of the Soil Conservation Service, Department of Agriculture.

All quoted material, except where noted otherwise in the text, involves personal letters to or from C. W. Thornthwaite found in the files of C. W. Thornthwaite Associates, Centerton, N.J., or in the personal files of the present authors. The former are reproduced here with the kind permission of William Superior, president of Thornthwaite Associates, and the Thornthwaite family, who also graciously supplied photographs used in this volume. We apologize if we have failed to acknowledge any source or to obtain needed permission, but it has been difficult to trace down individuals or their surviving family members after so many years.

Further, we would like to express our great appreciation to Joanne Danoff and Cyndi Timko in the Department of Geography, University of Delaware, and to Kathleen Lamothe of Waterloo, Ontario, Canada, who contributed so ably in typing the various versions of the manuscript. Their skill and patience have made the task of preparation both possible and enjoyable.

THE GENIUS OF C. WARREN THORNTHWAITE, CLIMATOLOGIST-GEOGRAPHER

Chapter 1

Early Years:
1899–1926

The Thornthwaite family name came originally from Cumberland County in the Lake District in northwestern England, a region controlled for many years by the Norse. *Thwaite* is a Norse term for field, and *thornthwaite* means a field cleared of thorns. Even today there is a village called Thornthwaite near the area thought to be originally settled by the Viking Honig. Cumberland records show an Adam de Thornthweyt in 1285. The name was spelled Thornweyt in 1543, Thornthat in 1605, and Thornthwaite in 1623. Warren Thornthwaite was very interested in his family history; he visited the Cumberland area on his first trip to England and sent postcards to his friends from the village of Thornthwaite.

Warren's paternal grandfather, William Thornthwaite, was born in 1848 in Nova Scotia in Canada. He was married in 1868 to Isabelle, who came from Maine and was just nineteen years old. The couple had five sons. The two oldest, Edward and Harry, were born in Nova Scotia, but the family moved to Michigan in 1873, before the three youngest—Calvin, Ernest, and Charles Warren—were born. The fourth son, Ernest, was to become Warren's father. William Thornthwaite was a farmer as well as a harness maker, and his family settled in central Michigan near Bay City.

In 1898 Ernest Thornthwaite married Mildred Hudson, who lived in Essexville, a few miles from Bay City. Mildred's father, Henry Hudson, who had been born in Elyria, Ohio, in 1846, was a farmer from Hampton Township. Her paternal grandfather, Joseph Hudson, had been born in Berkshire, England. Her paternal grandmother, Fidelia Essex Hudson, had lived in Connecticut and was a descendant of Ebeneezer Metcalf, who had served in the Revolutionary War in Colonel Experience Storr's Connecticut Regiment. The information on Mildred's ancestors comes from an application form for admission to the Daughters of the American Revolution that Mildred completed after Ernest's death and kept with her other important papers.

Joseph and Fidelia Hudson had been married in 1846 in Ohio, but one year later they moved to Bay County, Mich. At that time, Bay City contained only about a dozen houses, and the surrounding country had a large number of Native Americans (Indians, as the settlers called them). The Hudsons purchased from the government forty acres of land in Essexville, about three miles east of Bay City. The only ways into town were an Indian trail along the river and by canoe. Joseph and Fidelia had one son, Henry, and a daughter, Blanche. Henry lived on the home farm and was Warren's maternal grandfather.

After their wedding, Ernest and Mildred Thornthwaite lived on a farm near Pinconning in central Michigan, about twenty-five miles north of Bay City, where Ernest carried on his trade of harness making. This area of Michigan was just being settled, and for the Thornthwaites it was a pioneer existence (Slentz 1988). The pioneer families found that growing crops on newly cleared acreage took a great deal of time and energy. The trees had to be cut, and the stumps took up a good deal of space and made plowing and cultivating difficult. Plants near the stumps often did not grow well, and hand-operated implements had to be used. Corn and beans were planted one hill at a time with a hand-operated planter, and because of the stumps, a one-horse single-row cultivator was used. Peas and millet were also grown, the seeds spread broadcast by hand and then harrowed. Corn was harvested by cutting the stalks off near the ground, and these were put into shocks in the field. Later the ears would be husked and the fodder fed to the cows. Bean plants were loosened from the ground with a four-tine fork and later hulled and placed in small stacks for drying. With luck, the rain would not come until the small stacks could be moved into the barn, where they were flailed during the winter to separate the beans from the pods. Clearing wooded areas meant cutting or pulling up brush and trees and dragging large logs left by the loggers into piles and burning them. Logs and trees near the edges of the area were used for fencing. It was not an easy life for the farm families.

All four of Ernest and Mildred Thornthwaite's children were born on the farm. Charles Warren was the first, born on March 7, 1899, followed by Mildred in 1903, Fred in 1906, and Faith in 1914. Warren, as he was always called, early became familiar with the harness maker's trade. Later in life, he wrote to a colleague, A. Redfield of Woods Hole: "I come from a family of harness makers; my great-grandfather, grandfather, and my father were harness makers, and I had an early apprenticeship in the trade. Even yet, I regard the smell of leather as the best perfume in existence."

There were very few roads in that part of Michigan when Warren was growing up. Fred, the only sibling still living, remembers visiting Essexville and Bay City with his mother. They went three miles by horse and buggy to a railroad crossing, flagged the train with a white handkerchief, and then rode on the train to Bay City. They visited the Thornthwaite relatives in Bay City or stayed with the Hudson family in Essexville. By horse and buggy, it would have been a three-day trip, one way.

Most of the food needed by the pioneer family was grown on the farm, and all the children had to contribute. For the boys, there were cows to milk, chickens to feed, crops to attend to,

Charles Warren Thornthwaite in 1901 at age two.

and berries to pick. The girls helped their mother in the house with the washing and ironing, cleaning, and preserving. The little cash they could obtain was from the sale of eggs, or fruits, and vegetables in the neighboring towns. Clothing and any small luxuries were purchased through the Sears catalog. The Thornthwaites, like many of the enterprising settlers of Michigan, soon realized that education was the way out of the drudgery of farm life.

There was no high school in Pinconning, the nearest town,

Warren Thornthwaite *(second from left)* with his father, Ernest; sister Mildred; and mother, Mildred, about 1906.

so Warren went to live in Mount Pleasant when he graduated from grade school. He solved the problem of living away from home by being hired as janitor of the Methodist church and living in a room in the church. Warren graduated from high school in 1918, just before the end of World War I, and was inducted into the Army Cadet Corps. However, the war ended in November, the corps was disbanded, and Warren enrolled at Central Michigan Normal School (now Central Michigan University) in Mount Pleasant to earn the credentials to become a schoolteacher. He managed to support himself during this time with help from the Students' Army Training Corps.

It was at the normal school that Thornthwaite met Denzil Slentz, who was to become his wife. Denzil's family lived in Clare, not far from Mount Pleasant. Like Warren, Denzil was one of four children: a brother, Floyd Slentz, now lives in Walnut Creek, Calif,; one sister, Eunice Ann, died in 1980, and the other, Helen Irene, died in 1986. Denzil's background was very similar to Warren's.

Denzil's parents lived on a farm near Monticello, Ohio, where Floyd was born in November 1893, Denzil in 1899, and Eunice in 1901. In 1904 the Slentzes moved to Clare County, Mich., where they had bought a farm. The farm was only partly cleared of stumps and had an unpainted, one-story house. It was pioneer living, similar to that experienced by the Thornthwaites, and money was always scarce. Because the area had no school, in 1905 the two older children were sent to their grandparents' home in Ohio for two years for schooling, when Floyd was twelve and Denzil was six. After Denzil and Floyd had been two years in Ohio, their parents wrote that a school was opening near their home in Michigan and that they could return. That same year the third daughter, Helen, was born. Denzil's father decided to take his examination to be a rural schoolteacher and did this successfully in 1908. Floyd graduated from the local school that same year and was accepted at Central Michigan Normal School for a two-year course leading to a teaching certificate. Mount Pleasant was fifteen miles south of Clare and thirty miles from the Slentz home, so Floyd had to board in Mount Pleasant. Because of the cost of educating their children so far from home, the Slentzes decided in 1908 to move to Mount Pleasant. They rented a house, and Denzil's mother put a "Boarders Wanted" sign in the window. Business was good, and they had twenty boarders that first year. In 1911, they bought a house that was closer to the normal school and could accommodate forty boarders. It was called the Slentz Club.

Denzil also enrolled in the normal school, and during the winter of 1917 she contracted diphtheria. She recovered, resumed her schooling, and in 1920 graduated with a Life Certif-

Warren Thornthwaite and Denzil Slentz in Michigan in 1923, two years before their marriage.

icate with a specialization in teaching kindergarten. In 1923 both Denzil and Eunice obtained teaching jobs in Clare, Mich., fifteen miles from Mount Pleasant, and Denzil taught there until she married in 1925.

It was also at the normal school that Warren Thornthwaite met John Leighly, who was to have an important influence on his life. Leighly had come to the school in 1919 to obtain his high school teaching certificate. He and Warren met during the 1919 spring recess as members of a crew that unloaded lumber from boxcars; they soon became partners and did the work by themselves, efficiently and cheaply. Warren and John Leighly maintained their friendship until Warren's death in 1963. Leighly was to become an outstanding figure in geographic circles as a physical geographer at the prestigious University of California, Berkeley, where he had received his doctorate.

Leighly (1979) recalled that one of the requirements for the teacher's certificate at Central Michigan Normal School was a course in geography, and the professor of geography at the time was R. D. Calkins, who had done graduate work in geography and geology at the University of Chicago. He taught for a year at Ypsilanti (Michigan State Normal School) with Mark Jefferson and then moved to Mount Pleasant, where he remained until he retired. Calkins used the Socratic method in his teaching, which he had learned from R. D. Salisbury in Chicago, and this appealed to his young students, Leighly and Thornthwaite. Calkins taught courses in physical and historical geology and geomorphology, as well as a field course. Thornthwaite always felt he owed a debt of gratitude to R. D. Calkins, and when Calkins died in 1955, Warren wrote to Leighly, "I suppose you heard from Joe Carey that R. D. Calkins died in late September. It gave me an empty feeling. Now there are no ties left with Mount Pleasant."

After Leighly's first year in Mount Pleasant, Calkins urged him to go to the University of Michigan and helped him to obtain an assistantship in the Department of Geology and Geography through Carl Sauer and the head of the department, W. H. Hobbs. It was in this way that Leighly met Sauer, who was to be his friend and mentor for the rest of his life and with whom Thornthwaite also obtained his doctorate. Sauer, who had obtained his doctorate from the University of Chicago, had come to the University of Michigan in 1922 as an instructor in geography in the Department of Geology, which ultimately became the Department of Geology and Geography. He was very concerned about destructive land use in Michigan, and in 1922 he founded, under state auspices, the famous Michigan Land Economic Survey. In the spring of 1923, Sauer left Michigan to take an appointment as a professor at the University of California in Berkeley. Sauer offered Leighly the job of associate at Berkeley, which meant that he could pursue graduate work while teaching courses in cartography. This Leighly was pleased to do, and in 1927 he received his Ph.D., the first Berkeley doctorate in geography. Leighly remained at Berkeley until he retired.

John Leighly believed that the Central Michigan Normal School was a good place for Thornthwaite as an undergraduate because it offered a range of courses sufficient for the nonspecialist if chosen carefully. Thornthwaite's fondness for both mathematics and music dates from these years, when his favorite professors were Webster Pearce, who taught mathematics, and Harold Powers, who introduced him to the delights of music. Evidently Warren sang in the college chorus and took part in student performances of light opera. It was probably Professor Powers who introduced Warren to grand opera, which remained a passion throughout his life. He took several courses in geography from R. D. Calkins but, unlike Leighly, did not seem to favor that subject especially.

After graduating from the normal school in 1922, Warren obtained a job teaching science at the high school at Owosso, Mich., not far from Bay City, and his mother and Faith came to Owosso to keep house for him. It is unlikely that Thornthwaite taught any geography courses there, but during the summers of 1923 and 1924 he took courses at the University of Michigan, perhaps following John Leighly's advice. During these years as a teacher, he further developed his love of classical music and opera and even wrote and arranged a production called "A Night in Romany—A Gypsy Fantasy" for the Baptist Men's Entertainment Club. The production featured selections from Bizet's *Carmen* and Verdi's *Il Trovatore,* as well as songs by Robert Schumann.

In 1924, Warren was persuaded by John Leighly to resume his studies full time, and entered the University of California to do graduate work with Carl Sauer. Warren and John Leighly drove to California in a Model T Ford. It must have been quite a trip, because it was one of the early stories Warren related to his children. His daughter Elizabeth recalls him saying that there were few roads between Michigan and California at that time, so they often had to drive through the fields. He also maintained that, because the gas tank arrangement in the car was such that the gasoline couldn't feed while the car was going uphill, they had to back up the mountains to maintain the gas flow.

Because of the great influence of Sauer throughout Thornthwaite's life, it is important to examine Sauer's concept of geography. Of German background, the young Sauer had grown up in Missouri and had been sent to school in Germany, so he spoke and wrote German perfectly. This probably influenced both Thornthwaite and Leighly to read German and to become acquainted with German scientists, among them the noted climatologist Wladimir Köppen.

At Berkeley, Sauer taught an introductory course in physical geography, and it was important to his students Leighly and Thornthwaite that climate, rather than W. M. Davis's concepts of geomorphology, formed the nucleus of this course. Also, it appears that Sauer rejected the popular geographic concept, espoused by Ellen Churchill Semple, of the influences of the geographic environment on humans. Although probably best known for his work in cultural geography, Sauer also renewed a theme that had been forgotten in geography, that the human race was an agent in changing the face of the earth, as explained by George Perkins Marsh in his 1864 book, *Man and Nature*. Sauer was very fond of the word *stewardship* and often wrote of people's obligation to the earth, from which they derive their living. In this, he was decades ahead of his time, and it is only in the late twentieth century that the idea has gained popular acceptance. However, it was a concept firmly espoused by his young disciples, Leighly and Thornthwaite, and that came to fruition in a conference organized in 1955 on "Man's Role in Changing the Face of the Earth."

Early Years

As students of Carl Sauer, Thornthwaite and Leighly were fortunate in being exposed to visiting geographers whom Sauer invited to Berkeley. Albert Penck, the famous German geographer, visited in 1925, as did W. M. Davis, who was probably the most famous American geographer of the time. Davis was the first president of the Association of American Geographers, and a generation of geomorphologists subscribed to his ideas of the cycle of erosion. Other visitors were Michigan geographers Wellington Jones, Ken McMurry, Stanley Dodge, and Preston James. Leighly later stated that by far the liveliest of these fleeting visitors was Griffith Taylor, who stopped briefly on his way from Australia to Chicago and who, in 1935, inaugurated the first department of geography in Canada at the University of Toronto.

Physical geography was taught at Berkeley, although Sauer emphasized historical geography. However, as Leighly pointed out, the most important thing students learned at Berkeley was the sense of participating in the perennial enterprise of intellectual discovery, the great unexplored "ocean of truth."

In the late summer of 1925, Sauer's department was joined by students Fred Kniffen, Sam Dicken and Peveril Meigs. Kniffen was also a Michigander, having been a student with Leighly at the University of Michigan in 1919–1922, graduating in geology in 1922. Fred Kniffen, in a letter to one of the authors, remembers that early in those days they found out that Warren was fond of classical music:

> Sam Dicken and I got in the habit of buying him a classical record when he invited us to his place for a meal. Each time we had such an occasion, we three would kneel before a corner with the record ensconced on a globe. We knelt before the altar and besought the blessings of the Great God Sauer! And Warren played the records, sometimes bursting into the vocal parts with a really good voice that suggested some training.
>
> Warren would only rarely join Sam and me on our wilder

escapades, trips to drink a little Dago (red) wine in San Francisco. I think he really didn't enjoy such occasions, the drinking that is. I was in the field a great deal during my four years at Berkeley. Warren went with Pev Meigs to Baja California the summer before my arrival. Much of it was mapping major geologic features about which neither knew much. I don't recall that Warren was a participant in the field courses that were part of the general requirements. Sauer gave the course, but physical geography was not Carl's forte.

In 1925, Warren asked Denzil to marry him, and the wedding was to take place in California. Denzil's family learned that a neighbor, Mrs. Lee, was going by train to Oregon to visit some of her children that summer, and she agreed to have Denzil go with her and stay with her family until it was Denzil's time to go to Berkeley. When Denzil arrived in Berkeley, she found that Warren had arranged with a friend to have the wedding at his house. This friend had a cabin somewhere that the newlyweds used for a honeymoon. Their daughter Elizabeth recalls being told about the events of this period:

My mother took the train to California after my father had gotten established there. At that time, for a young lady (she was 25 years old) to go by train from Michigan to the West Coast was quite an adventure. She wrote in one of her letters home that she saw a strange plant growing everywhere. It turned out to be a palm tree, which of course she had never seen. They were married there in Berkeley at the home of "Uncle Foster" (a friend, not a true uncle). They drove around California on their honeymoon and spent one night (at least one) in a haystack somewhere.

My mother contracted tuberculosis after they were married and spent a year in a sanatorium. My father often commented that it was difficult to be a graduate student AND have a wife in a sanatorium. (personal communication to Sanderson, 1990)

When Denzil contracted tuberculosis, her sister Eunice went to Berkeley to help care for her. Denzil had improved considerably by November, and Eunice returned home only to find that she too had tuberculosis. She spent most of the winter in bed. Warren finished his courses at Berkeley in June 1927, and he and Denzil went to Michigan to see the family. Denzil found she was pregnant, so when they left for Oklahoma in September to begin full-time teaching, Eunice went with them to help them settle and care for the baby.

Professor George Carter, who later became chair of the Johns Hopkins University Isaiah Bowman School of Geography, followed Thornthwaite to Berkeley about ten years later. Carter had many opportunities to know Thornthwaite well and recalls many stories of Thornthwaite and Leighly at Berkeley, stories that were still circulating a decade after both had graduated:

Warren Thornthwaite was a legend when I went to Berkeley for graduate study in geography in 1938. He roomed with John Leighly, and they walked to and from the campus. Warren was known to have said that when John was gestating an article, days on end went by without John speaking one word with him. He walked in silence, but once he started writing, he simply wrote out the whole article without a word changed.

Sauer sent Warren and at least two others into the Southwest. I don't know just who the group was. Sauer told me that they quarreled and that in hopes of quieting them down he sent the wives into the field with them. It only made things worse. Sauer obviously was not too astute a student of human nature for when it comes to their families, women are the ultimate partisans.

Money was a problem for the young Thornthwaite, and in 1925 he began to work part-time with the Kentucky Geological Survey in Frankfurt, Ky. Because of Thornthwaite's need to be in Kentucky, Carl Sauer, who at that time was becoming interested in the geography of cities, suggested that he undertake research on the city of Louisville for his doctoral dissertation.

Thornthwaite was awarded his doctorate in 1930. His dissertation was entitled "Louisville, Kentucky: A Study in Urban Geography," and a copy is preserved in the files of the

University of California, Berkeley. The dissertation is certainly unlike anything else written by Thornthwaite, perhaps reflecting Sauer's rather than Thornthwaite's research interests. Probably it was also a practical choice, considering that Louisville was close to the area where he was working. Reflecting some of Carl Sauer's ideas, it deals with architectural types, urban structure, and functional differentiation in Louisville. It is some 200 pages long and contains 140 figures, the latter reflecting Thornthwaite's interest in maps and diagrams. In the acknowledgments, Thornthwaite stated that he did field work in the area during three summers, the first assisted by his brother Fred, the second by his colleague Sam Dicken, and the third by his wife, Denzil. It may have been as much Denzil's as Warren's dissertation! Thornthwaite's interest in climatology is evident only in an appendix entitled "Climate of Louisville," in which he stated that it belonged to the Cfa climate zone of Köppen (see chapter 2).

In one section, Thornthwaite's sense of humor is evident. He noted inscriptions on tombstones in Louisville and quoted one that he saw repeatedly: "Remember friend as you pass by / As you are now, so once was I / As I am now, so must you be / Prepare for death and follow me."

Warren stated that on one stone there was penciled this reply: "To follow you I'm not content / Until I know which way you went."

Chapter 2

The University of Oklahoma: 1927–1934

When Thornthwaite obtained a position in 1927 as a professor of geography at the University of Oklahoma in Norman, he and Denzil moved there from California. Elizabeth, the first of the three Thornthwaite daughters, was born on November 17, 1927, in Oklahoma City because there was no hospital in Norman at the time. It was depression time when Elizabeth began school, and she remembers being envious of her classmates who came to school barefoot. However, her mother told her, "You have shoes, and you will wear them!"

In 1927, the University of Oklahoma had a student body of about six thousand. The university had graduated the first class (of four) in 1898. In 1921–1922 the Department of Geology, then in the School of Mines, was renamed the Department of Geology and Geography; in 1923 the geography courses were listed separately and a major program in geography was announced. The Department of Geography was transferred in 1927 to the College of Arts and Science. Thornthwaite joined C. J. Bollinger as the second member of the department. Bollinger was another Michigander who had come to Oklahoma in 1920. Ruel B. Frost, who later achieved national recognition as head of the Geology and Geography Department at Oberlin College, and John L. Page, who went on for a doctorate at Clark and taught at the University of Illinois, were the first two graduates in geography at the University of Oklahoma, receiving their bachelor's degrees in 1926.

At that time, geography was classified in the "mathematical and natural sciences" group of the College of Arts and Sciences, into which geology and geography had been transferred upon dissolution of the School of Mines. Enrollment in courses above the survey level was extremely small—often fewer than five students in those early years—but the few students majoring in geography were inspired by the vigorous teaching and research activities of the two-man staff to pursue advanced studies themselves. The first master's degrees were granted in 1930. The 1930–1931 catalog of the university stated, "The objectives of the department of geography are as follows: provide instruction for students interested in the subject as a part of general education; provide a geographical background for students of economics, history, and political science; meet the needs of students desiring to specialize in geography either for teaching or commercial work; and encourage and conduct researches particularly in the geography of Oklahoma and the Southwest."

In the archives of the University of Oklahoma can be found a listing, compiled by John Caldwell, of the courses taught by Thornthwaite from 1927 through 1934. It seems incredible that one person could teach so many different courses! An example is given below:

Fall 1927

Geography 1—World Regional: "Theoretical basis of regional classification and a systematic regional survey of the continents."

Geography 6—Physical Geography: "The physical features and phenomena of the earth's surface, their causes and consequences. Emphasis on those features of the physical environment which are of greatest significance to man."

Geography 8—Economic Geography: "Principles of Economic Geography."

Geography 127—Geography of North America: "A regional survey of the physical and economic geography of the United States (including Alaska) and Canada."

Geography 250—Seminar in Geography (with Bollinger)

Spring 1928
Geography 1—General Geography: "A study of the distribution, interaction, and association of the natural and cultural phenomena of the earth's surface."
Geography 110—Cartography: "Elements of map making, map projections, conversion of scale, technique of representing geographical data, development of facility in choosing projections to fit particular needs, exercises in field mapping."
Geography 125—Industrial Geography: "An interpretation of the distribution and localization of mining and manufacturing with special consideration of development of cities and allied cultural forms. The increased dependence upon minerals with the advance of culture. The limited supply of certain minerals, and the need for conservation. The position of the various nations with respect to mineral production and reserves."
Geography 130—Geography of Asia: "Special emphasis on the population problem."

Each semester, until spring 1934, Thornthwaite taught four or five courses, adding new topics such as urban geography, geography of Oklahoma, geography of Africa and Australia, and principles of human geography.

In those early days in Norman, the Thornthwaites, with daughter Elizabeth, spent their summers in Michigan with the relatives, and in 1929 they invited Denzil's young sister Helen to return to Oklahoma with them and pursue a master's degree at the university. This she did, and she later collaborated with Warren in some of his research. It is quite likely that Thornthwaite took advantage of these summer periods to join a group of other geographers, naturalists, and foresters on camping surveys through southern Canada and as far north as the James Bay region. Later in his career, Thornthwaite recounted how he had bored holes in thousands of trees on his Canadian expeditions. There is no record of what he did with the tree ring data he collected, but there is reason to believe that he was looking for evidence of year-to-year climatic variations and information on how the temperature of each year revealed itself in tree growth.

While at Norman, Thornthwaite kept in touch with Carl Sauer in California, and there is some Sauer-Thornthwaite correspondence in the Sauer archives in Berkeley. Evidently, in 1932 Sauer attempted to obtain a position for Thornthwaite at Harvard and also tried to obtain a publisher for Thornthwaite's thesis. Unfortunately, neither effort was successful.

Arnold Court (a professor emeritus of California State University, Northridge) recalls that in 1933, he transferred as a junior into the geography program at Norman. Geography then occupied both floors of an old wooden building, formerly the gymnasium. Classrooms were downstairs and laboratories and offices upstairs. In addition to beginning courses in physical geography with Bollinger, and regional geography with Thornthwaite, Court took a course in economic geography with Leslie Hewes, who had joined the department in 1928. Court recalls,

> Bollinger was deeply immersed in crop climatology, and had begun his life-long research on extra terrestrial (solar and planetary) factors in climate, especially Great Plains summer rainfall and corn yield. Thornthwaite sought better ways of computing boundaries for the first version of his new climatic classification which had been published two years earlier. In those three beginning courses, Bollinger taught the Köppen climatic classification, Thornthwaite his own, and Hewes a version of the British scheme used by Kendrew and Miller. I developed an abiding dislike for all rigorously defined climatic types and especially their boundaries.

Thornthwaite taught from "Regional Geography of the World, Part I, The Old World (The lands occupied by advanced and powerful cultures prior to the age of the great discoveries) by C. W. Thornthwaite, University of Oklahoma, 1929" [unpublished class lecture notes]. The 70 mimeographed pages were double-spaced outlines, except for single-spaced pages of "Selected Readings," four on seven pages for Chapter I (Scope of Regional Geography) and three on eleven pages for Chapter II (Origin and Spread of Mankind and Culture). For "The European World," outlines covered successive chapters on the Nordic, Romanic, and Slavic Realms, the Shatter

Belt, and the Levantine, Oriental, and African Realms. At the back was a five page "Outline of Geography 2" (perhaps the Berkeley version; this was for O. U. Geog. 43), and a page and a half of "Selected References."

Course homework consisted in drawing, on outline maps of the world, Europe, or parts thereof, various cultural and economic aspects: culture regions of the world (Goode homolosine); languages of Europe; vegetation (natural and agricultural) of Scandinavia; iron, coal, canals, railroads, and cities of Germany; Greek and Phoenician settlements in the Mediterranean Basin; the Roman Empire ca. 395 A.D.; ancient French provinces and wine and cider production areas; coal and iron in France, with orography; and a linear graph of England-Wales population since the Norman conquest and a column graph of land use in each of 12 countries.

Thornthwaite explained at the first meeting of about 40 that each graph or map would be worth either 10 or 20 points, and that course grades would depend on totals of such points plus those earned on quizzes and examinations. Scoring of maps would depend on both accuracy and neatness; a perfect score could be attained only by careful choice and application of suitable and harmonious colors with special water-soluble pencils, washed with a water dipped brush to attain pastel effect. Following the same numerical philosophy which caused much confusion in his climatic classification, an A would be granted the student with the most points and all those with ⅞ or more; B to those with ¾ to ⅞ of the leader, C to ⅝ to ¾, D to ½ to ⅝, and F to any with less than half the leader's total.

To maintain confidentiality, Thornthwaite assigned to each person on the class roster a secret number, and posted cumulative scores by such numbers (students had no computer ID numbers then). But roll was called every day, and within a few weeks the class roster could be compiled by any interested student. The numbering was found to begin with Mitchell and proceed backward through the alphabet to Moore.

At mid-term, posted scores showed five A students, with a gap of more than 20 points above the rest of the class. As three of the five leaders were coloring assiduously on the next map, one commented in the lab that the instructor's insistence on coloring skill was more appropriate to the first grade than a non-art class in college. Another pointed out that the A to B gap was more than the score for any one map, and that if all five agreed to forego one map the final score distribution would be less discontinuous. Two students had already identified, from the unofficial class roster, the other three leaders; all five solemnly agreed to skip the next map.

When Thornthwaite handed back the submitted copies of that map, he told the class, in a very pained voice, that the five best students had conspired to wreck his grading system, and that consequently he would revise the rules. Grades would be based, not on the highest score, but on the total possible, with the strong possibility of no A's at all and more failures. He said that the assignments, and his methods of scoring, were carefully designed for instructional value, not as busy work, and he was very disappointed at the crass attitude of his best students. Nonetheless, each of the five renegades received an A!

It is incredible that Arnold Court should remember, after all these years, such details about Thornthwaite's marking system! Court also remembers some personal Thornthwaite anecdotes:

For several years, Denzil and Warren Thornthwaite had rented a large two-story house just a block from that of my parents, where I was still living while attending college in those depression years. Often Thornthwaite and I would walk together the half-mile from the Geography Department to our houses, and on several occasions I entertained Elizabeth Thornthwaite with stories, and especially verses from Lewis Carroll: "Twas brillig and the slithy toves . . ."

During the spring semester of 1934, Thornthwaite went East for undisclosed reasons, which resulted in his taking leave the following year to work on a migration study at the University of Pennsylvania. Never again would he return to full-time teaching, though he maintained academic connections for most of his life.

Another student of Thornthwaite's in Oklahoma who later became a colleague was Leslie Hewes. Thornthwaite was Hewes's professor for two or three semesters and influenced the young geographer to do graduate work with Carl Sauer at Berkeley. Hewes recalls,

Thornthwaite was pivotal in my decision to do graduate work at Berkeley. In his opinion, Sauer had the true religion and was the leading exponent of what geography was about. Prior to my returning to the University of Oklahoma as Instructor in Geography, I substituted for Thornthwaite in teaching a small class in a cross-country bus tour the previous summer. He had signed up to teach the class, but decided to do something else. The class was on the geography of the western part of the country and was accepted for credit at the University of Oklahoma.

When returning to teach geography at Oklahoma in a combined Department of Geology and Geography, I found myself in an awkward position between Thornthwaite and the senior geographer (Bollinger) as the go-between between two men who were not speaking. I was never sure of the cause of the friction, whether because of the fact that Thornthwaite had his Ph.D. and the other man did not, or whether it had to do with Thornthwaite's system of climatic classification, or something else. At any rate, I became aware that Thornthwaite was very sensitive about his climatic classification. I suppose that friction in the office was a factor in his decision to leave the department to take part in the study of population change and then to go into government work.

Thornthwaite worked part-time for the Kentucky Geological Survey in Frankfurt, Ky., during the years 1925–1930, first while he was still a student at Berkeley and, after 1927, while he was a professor at the University of Oklahoma. From a bibliography of Thornthwaite's writings, it is noted that no published article resulted from this work. Thornthwaite's first publications were by the University of California Press: a base map of California in 1926 and one of Eurasia and Africa in 1927.

Thornthwaite's first article dealing with climate was "The Polar Front in the Interpretation and Prediction of Oklahoma Weather" in 1929, in the *Proceedings of the Oklahoma Academy of Science*. It is probable that he was intrigued with the climate and weather in Oklahoma, which was very different from the climates he had experienced in California and Michigan. He began his long association with the *Geographical Review* in 1928 when he wrote "A Reconstruction of the Natural Vegetation of Ohio," a review of two articles by Paul Sears that had been published in the *Ohio Journal of Science*.

Thornthwaite's first article that attracted worldwide attention, published in 1931, dealt with climate classification. Thornthwaite had first become interested in the classification of climate while studying with Sauer in California, where he read the work of the German climatologist Köppen in the original German. During his first years at the University of Oklahoma, he did a good deal of research on the work of Köppen, whose classification of climate, originally published in 1900, was already universally known and accepted. His colleague at Oklahoma, Bollinger, taught the Köppen system of classification, and the observant Thornthwaite did not fail to see how the classification fell short in describing Oklahoma's climate.

Before attempting an examination of Thornthwaite's 1931 classification, it is important to look at the life of Wladimir Köppen, the climatologist who had a great impact on Thornthwaite and his writings. The following material on Köppen's life was kindly provided by Rudolf Geiger, who was a young colleague of Köppen's during the latter part of his life:

Wladimir Köppen was born September 25, 1846 at St. Petersburg in Russia. His grandfather Johann Friedrich Köppen was a German physician who had been invited by Catharine II of Russia to help organize that country's Public Health Organization in Russia. So W. Köppen was born in Russia although remaining a German citizen. He was the youngest of six brothers and sisters, and, until he was 13 years old was taught at home by two older sisters. He spent 1859 and 1860 in a Russian Gymnasium in St. Petersburg and in 1864 he began to study at St. Petersburg University. In 1867, he went to Heidelberg University in Germany and then in 1869 to Leipzig University where he obtained his doctorate.

From 1872 to 1873 he was an assistant at the Central Observatory in St. Petersburg, but in May, 1875, he returned to Germany to the naval observatory in Hamburg, where his job was manager of the weather division. In 1876, he published the first synoptic weather map of Europe. He supervised the installation of observation-stations along the coastline of the North Sea and was the first to organize a system for the prediction of storms. He was married on November 15, 1876 to his "Marie," who was his dear wife till her death on February 16, 1939, eighteen months before his own death. He lost three sons during the 1914–1918 war, but at least two daughters survived him.

In 1879 he was appointed Meteorologist of the German Naval Observatory in Hamburg, his duties being to help in all types of scientific work, and he remained there until April 1, 1919. In 1892 he was editor of the "Annalen der Hydrographie und Maritimen Meteorologie," and in 1893 he was the first person to obtain upper air information by using a kite. In 1903 he installed a German upper air station at Großborstel near Hamburg, thus becoming the father of upper air investigations in Germany. In 1924 he emigrated to Graz in Austria, where his son-in-law Alfred Wegener (of continental drift fame) was living. As a very old man, Köppen was the meteorologist who initiated the meteorological navigation for the first commercial airplane flights across the Atlantic Ocean. He died at Graz on June 22, 1940. (personal communication to Sanderson, 1972)

In Köppen's youth, the science of botany was rapidly developing. Alexander von Humboldt was at the height of his career, and in 1859 Charles Darwin published his *Origin of Species*. The German botanist, H.W. Dove published the first world maps of monthly temperature and precipitation in 1848, and in 1866 A. Grisebach published the first world map of vegetation. As a student, Köppen's interests were in botany and especially in phenology, and his doctoral dissertation at Leipzig University in 1871 was on the subject of heat and plant growth. Köppen's major writings in climatology were done while he was in Hamburg, where in 1884 he published his first article on climate, which divided the earth into temperature belts. Köppen's early training in plant physiology and his familiarity with Grisebach's work on plant geography led him to the idea that plants might serve as integrators of the climatic elements.

After searching for a plant classification to use, Köppen chose the classification of A. de Candolle, which used the symbols A, B, C, D, and E to represent the five vegetation groups, with "A" representing the plants of the macrothermal or tropical areas (the torrid zone of the Greeks). The "B" and "C" plants shared the "temperate zone" of the Greeks, with "D" representing the microthermal and "E" the frigid zone. The B group represented the dry or neophytic plants.

Köppen spent a great deal of time developing numerical climatic formulas for defining de Candolle's vegetation boundaries. His first important paper on climatic classification appeared in 1900, and the general terms of the classification are familiar to all geographers. The first letter in the classification refers to the temperature regime of the place (e.g., the "A" climates are the tropical climates, in which no month has a mean temperature of less than 18°C). The second letter of the classification refers to the precipitation regime—"f" (*feuchte*, or moist, in German) means no dry season, "s" means dry summer, "w" means dry winter, and "m" means monsoon or summer rain. The classification was a great step forward at the time of its origin, producing a numerical system, based on actual measured data, of the world's climates. It did, of course, have its drawbacks. It was based on the very tentative map of world vegetation by de Candolle. Since that time many improved maps of world vegetation have appeared. And, of course, Köppen's climatic data were few and quickly outdated, so his attempts to fit climatic isolines to the vegetation boundaries were not always confirmed by actual observations. The classification worked best in Europe, where more data were available and also where moisture is usually not a limiting factor.

The definition of dry climates was always a problem for Köppen. In his 1900 study, he experimented with a variety of definitions in an attempt to arrive at an expression for effective precipitation. At first he placed the boundary between desert and steppe where the rainiest month of the year had an average of six days with rain or a probability of rain of 0.20. He also experimented with ratios of monthly rainfall to the saturation vapor pressure corresponding to the mean temperature of the month. He was, of course, attempting to arrive at a connection between rainfall amounts and the evaporative power of the atmosphere, but with the lack of data, this was impossible. In 1918, he compromised by using a simple ratio between mean annual precipitation and mean annual temperature. He realized the shortcomings of this method when he stated that precipitation alone is of little help in distinguishing between a moist and a dry climate.

Köppen contributed the section on the geographical distribution of climatic types to the famous *Handbuch der Klimatologie,* published in 1936. His papers were, of course, in German, and Köppen became known in North America only after English translations of his work appeared. It was the first numerical classification of climate based on actual observed values of temperature and precipitation. The English-speaking world eagerly adopted Köppen's classification, and it continues to be the most used classification to the present day.

However, as Thornthwaite pointed out, Köppen never regarded his classification as a finished product and often remarked on the need for a more rational foundation. He realized that his omission of a moisture effectiveness factor in his classification was a problem, and he realized that equal amounts of annual rainfall produce both forest vegetation in Siberia and desert plants in Africa. Köppen also realized that he had no climatic category termed "subhumid," and it was this fact more than any other that prompted Thornthwaite to recognize the shortcomings of the Köppen classification.

Thornthwaite was living in Oklahoma during the dust bowl days of the early 1930s, and he knew that the climate of the Oklahoma Panhandle was not humid as the Köppen classification stated. Neither was it semiarid; it had a unique climate, as Thornthwaite realized, transitional between humid and semiarid, which he classified as "subhumid."

"The Climates of North America According to a New Classification" required a great deal of research on Thornthwaite's part. He was thoroughly familiar with Köppen's publications, having read them in the original German, and agreed with Köppen that plants were a good index of climate. The Köppen classification brought out the general relationships of vegetation and soils to climate. Thornthwaite also knew of the attempts by various American climatologists to apply the classification to North America: E. McDougall's "The Moisture Belts of North America," W. Van Royen's "The Climatic Regions of North America," and R. J. Russell's "Climates of California." All agreed that there were inadequacies in the Köppen system and that there was a special problem with the evaporation term.

Thornthwaite clearly stated in the 1931 paper that it was the "effectiveness" of temperature and precipitation, rather than the crude measurement of these two parameters, that was important in climate classification. By "effective temperature" he meant the rate of plant growth that resulted from temperature, whereas the "effectiveness" of precipitation depended not only on the amount of precipitation but also on the amount of water lost by evaporation.

Thornthwaite then set out to try to determine precipitation effectiveness by using some measurements of precipitation and evaporation (published in the Weather Bureau's Bulletin W) to obtain what he called a "P-E index" for the place in question. Using the measures of evaporation for April to September at twenty-one stations in the United States, Thornthwaite ascertained the relationship between precipitation and evaporation and arranged these into five groups according to the average monthly temperature in degrees Fahrenheit. Thus he obtained an expression that permitted air temperature to be used in place of evaporation. He computed the value for single months, then summed the twelve months to obtain a "precipitation effectiveness index." Because computing the P-E index by means of the formula was a laborious process, Thornthwaite constructed a nomogram, which made the calculation simpler. By means of a comparison with vegetation, he established five grades of humidity: wet, humid, subhumid, semiarid, and arid. These he identified with characteristic vegetation as follows: rainforest, forest, grassland, steppe, and desert, respectively. Like Köppen, Thornthwaite gave letter names to the five humidity provinces. An "A" climate represented rainforest vegetation, "B" forest climate, "C" subhumid grassland climate, "D" semiarid steppe climate, and "E" arid or desert climate.

Thornthwaite also attempted to derive a temperature effectiveness (T-E) index, and this was done by purely empirical means. An equation was derived that gave the poleward limit of the tundra a T-E index of zero and the poleward limit of the tropical rainforest an index of 128. Six temperature zones were thus identified: A' (tropical), B' (mesothermal), C' (microthermal), D' (taiga), E' (tundra), and F' (frost). Thornthwaite proceeded to classify all the climatic stations in North America and to draw a map of the climates. The classification was very cumbersome, and the system was not much used. But it was, as he suggested, better than Köppen's for detailed analysis.

In Germany, Köppen, of course, read of the American modifications of his classification and of Thornthwaite's new classification, and it is interesting to read his reaction. In Thornthwaite's files there is a letter to him in German from Köppen, translated as follows by one of the authors:

Graz. 25 April, 1934

My Dear Colleague:
Some time ago I sent to you my only copy of the little map of the Köppen climates in the U.S.A. that is intended for the *Handbuch*. I

ask now for its return so that I may add something in two places. I fear I have forgotten the dry area of the Columbia River, and I see that in the western part of Utah a BW area must be added.

About two years ago, I received two fine works of classification of climates by Mr. C. W. Thornthwaite in Oklahoma and Mr. R. J. Russell in California in which something different was proposed, particularly for my B climates (dry class). That my very simple classification is capable of being improved, naturally I understand, and I would listen to the proposals with pleasure if their application to the whole world were proven. But for the *Handbuch* I can't do this, since then the compatibility of the continents with one another, which I would like to maintain above all, would disappear. The homogeneity appears to me the most interesting, something that, before me, scarcely anyone had really considered. But in time someone will work through these questions for the *whole* world, and then these American preliminary works will be an important help.

With best wishes,
Yours truly,
W. Köppen

In 1934 Köppen wrote, in German, an article entitled "Attempts to Perfect This Classification" which, translated as follows, is preserved in the Thornthwaite files:

In the last years in North America more works have appeared concerning the classification of climate. After 1919 R. de C. Ward and in 1927 J. B. Leighly have acquainted their countrymen with my endeavors. R. J. Russell has elaborated the climates of the United States in 1926 and 1932, and C. W. Thornthwaite in 1931 and 1933 the climates of the United States and the entire world with new presentations in this direction, which we must comment upon here. All four papers are accompanied by maps.

In his work on California, Russell has added three new climates to my map as subdivisions of it. Csn on a coastal strip about 6 kilometers wide, stretching north from Esteros Point, and a similar cloudy climate, BWn at San Diego, likewise the climate BWhh in the Colorado desert with the average temperature maxima of over 38°C in three months (Climate BWn corresponds to my Garua-Climate along other coasts, and the clouds in it and in the Csn strip appear to be just as saturated as at Lima and Cape Town).

In his second work Russell performs a very worthwhile investigation concerning the application of my older and newer definitions of BW and BS climates in the arid regions of the United States, which he has learned to know personally through countless trips. He finds, upon examination of the entire number of meteorological stations, that my definitions of 1922 correspond better to the vegetation conditions than the newer ones of 1928, which form the basis of this handbook. If, as is very possible, this is also confirmed in other parts of the world, then it would be advisable to change the definitions again to conform perhaps to a midpoint between these two.

For the C/D boundaries Russell prefers the January temperature of $0°$ to that chosen by me ($-3°$) and wishes to retain this also as the boundary between hot and cold arid climates, for which I propose a yearly temperature of $18°$. I prefer my boundaries on the following grounds: a normal winter snow cover does not appear at a January temperature of less than $0°$, but appears first at about $-3°$, as Penck also states, and therefore, for arid climates (except oases) I consider the finer classification according to the temperature of the coldest and warmest months on the whole superfluous; I choose instead a yearly temperature.

C. W. Thornthwaite studied these questions still more thoroughly in his two papers. Next he also proceeds from my classification; he follows closely my line of thought and thereby arrives at so complicated a formula that the general application of it is indeed scarcely to be expected, especially since the main outlines of the climatic system are almost spoiled. Yet the attempt to find a more rational foundation for the classification is, in all events, very worthy.

Finally, he finds it necessary, just as I did, to take into consideration the yearly precipitation, but unlike me, he does not coalesce these three different elements into chief types and subdivisions, but develops them independently of each other and designates the first two by equivalent symbols—theoretically correct, but practically leading to the loss of the whole perspective.

Immediately following the publication of his 1931 climatic classification, Thornthwaite was busy applying the classification on a worldwide basis. In total, Thornthwaite used climatic data from more than four thousand world stations, applying his criteria for classifying each station as A, B, C, D, E, etcetera, and drawing a world map. Although this map of world climates was more rational than Köppen's, it was very complicated to use and never became very popular.

In 1934, Thornthwaite published a review of some Australian climatic articles by Andrews and Maze in the *Geographical Review*. He was critical of these articles, stating that as there were so few climatic stations in Australia, there could be no proper quantitative climatic classification. Also in 1934, Thornthwaite reviewed a 1933 Japanese article by Isozaki on "Thornthwaite's New Classification of Climate and its Application to the Climate of Japan" (*Journal of Geography, Tokyo*) in the *Geographical Review*. In this article, Thornthwaite criticized the author for attempting to use the Thornthwaite P-E index to determine specific evaporation values in Japan. Obviously this cannot be done, as Thornthwaite pointed out, and the author's classification of the climates of Japan according to the Thornthwaite system is invalid.

While living in Oklahoma, Thornthwaite, as a true geographer and student of Carl Sauer, began to be interested in the economic and cultural development of the Great Plains region. He was invited by the *Geographical Review* to review three books on the Great Plains—*The Great Plains*, by W. P. Webb (Boston: Ginn and Co., 1931); *The Range Cattle Industry*, by E. E. Dale (Norman: University of Oklahoma Press, 1930); and *The Day of the Cattleman*, by E. S. Osgood (Minneapolis: University of Minnesota Press, 1929)—and the review appeared in 1932. Naturally the climatic problems of the Great Plains interested Thornthwaite, and he agreed with Webb that limited moisture availability was central to the question of settlement of the Great Plains. The Great Plains was to be the topic of an excellent, if little known, chapter by Thornthwaite in a

1936 book called *Migration and Economic Opportunity* published by the University of Pennsylvania Press.

Thornthwaite was an avid reader who read beyond his immediate scientific endeavors, and it appears that the *Geographical Review* was always sending him books or articles to review. In 1932, the *Review* published an article by Thornthwaite entitled "New Light on Climatic Change." In this article, he reviewed the progress made by botanists in pollen analysis and outlined the conclusions that the climate in North America had become warmer and more humid since the last Ice Age.

In the same year, the *Review* also published an article by Thornthwaite entitled "The Quantitative Determination of Climate." In this article Thornthwaite examined various climatic articles, including Russell's "Dry Climates of the United States" (1931), Köppen's "Die Klimate der Erde" (1923), and Geiger and Zierl's "Köppen's Klimazonen und die Vegetationzonen von Africa" (1931). Thornthwaite concluded that Russell's study was probably the most scholarly of the studies that had applied Köppen's classification to regions other than Europe. Thornthwaite criticized the Geiger and Zierl article for using Africa rather than North America as their test case, for, as Thornthwaite stated, Africa had only 328 climatic stations, whereas North America had at least four thousand, and the vegetation map of North America was certainly better than that of Africa. Thornthwaite concluded, "They must have known in advance that Köppen's system is in closer agreement with the vegetation zones of Africa than with those of any other continent" (p. 325). One wonders if, later in life, when Thornthwaite and Geiger became good friends, this criticism was ever mentioned.

Chapter 3

Washington and the Soil Conservation Service: 1934–1946

In 1934, Thornthwaite resigned from the University of Oklahoma to work with the Study of Population Distribution of the University of Pennsylvania. Although Thornthwaite never clearly revealed the real reason for this significant career move, it was known that he was restive because his heavy teaching load precluded the development of more interesting lines of climatic research. Even though teaching was an ongoing interest during his whole life, he was less enthusiastic about formal classroom approaches to instruction. Rather, he espoused a less-structured approach in which he could instruct by example, by individual one-on-one discussions, and in the daily interactions involved in carrying out research studies. There is also some suggestion that Bollinger and Thornthwaite did not always see eye to eye. Both were working in the area of climatology and had somewhat conflicting viewpoints. A move from Oklahoma would ease any tension, and this opportunity to work in Washington with the University of Pennsylvania on population distribution offered an excellent opportunity. The study was organized at the suggestion of the Social Science Research Council, with funds from the Rockefeller Foundation, to consider what movements of population would be necessary or desirable and what part the government should take in encouraging such movement. The reason for the study was, of course, the depression; with millions of people out of work, the question to be answered was whether the government should move people to where jobs might be found. Although not at all like his climatic publications, Thornthwaite's contributions to this study were of great value.

In 1934, his paper entitled "Internal Migration in the United States" was published. It was a thorough study with many excellent maps of interstate (and even county) migration of native white and black populations. The publication certainly indicated Thornthwaite's ability to deal with masses of data and to produce useful maps and graphs to illustrate the points of his paper.

In 1936 the University of Pennsylvania Press published the book *Migration and Economic Opportunity,* which contained a chapter on the Great Plains written by Thornthwaite. Thornthwaite mentioned that Isaiah Bowman had critically read the manuscript, and he acknowledged the assistance of O. E. Baker, who was then working in the Department of Agriculture.

His chapter on the Great Plains represents Thornthwaite's only article dealing with economics and population. He stated in his introduction that the Great Plains "is an area where climatic conditions unfavorable to a permanent agricultural economy recur with irregular persistency." Included with the article was a series of maps showing the various climates that occurred in the Great Plains for the years 1910–1934. It was the first approach to delimiting areas of climatic hazard in the Great Plains. He correctly identified the area as semiarid and concluded that only two alternatives exist for inhabitants of the Great Plains—permanent poverty or permanent subsidy.

He also mentioned a controversial government policy with which he disagreed: "In 1873, the Timber Culture Act was passed on the assumption that if the western farmers were induced to plant trees, rainfall would be increased to such an extent that climate hazards to crop production would be removed. The Act gave an additional quarter sector of land to the homesteader who would plant and maintain 40 acres of it in timber." (p. 209)

The Thornthwaites had moved to Washington in 1934, and Warren was working on his population papers at the Brookings Institute. The year before, in 1933, Roosevelt's New Deal government had set up the Soil Erosion Service in the Department of the Interior, with H. H. Bennett as director and W. Lowdermilk as vice director. At that time, Isaiah Bowman, president of the Johns Hopkins University in Baltimore, was chair of the Land Use Committee of the Science Advisory Board set up by Secretary of the Interior Harold Ickes. In March 1934, the Land Use Committee engaged Carl Sauer to prepare specific recommendations for the expansion of research on the relations of surface soil and climate to erosion. The committee recommended that a "Climatic and Physiographic Section" be established, and it was Sauer who recommended Thornthwaite as director of this division. It is known that Thornthwaite met Bennett, the director of the Soil Erosion Service, sometime in 1934, as the files of the service in the National Archives in Washington indicate that Warren was invited to a staff luncheon on September 20 of that year.

The Soil Erosion Service was not to have a long life; in October 1934 there were memos to Lowdermilk about a Soil Erosion Service Reorganization Plan, and in March 1935, Lowdermilk heard from the Secretary of the Interior that it was possible that the service would be moved to the Department of Agriculture. By April 23, this move had been effected. It appears that negotiations between Lowdermilk and Thornthwaite concerning the latter's employment by the newly named Soil Conservation Service of the Department of Agriculture were carried on during April 1935, and a memo from Lowdermilk to Thornthwaite dated May 4 stated,

Reference is made to your letter and memorandum of April 2. I have read over the outline of climatic studies and find that they are broad enough to include investigations that will be necessary for studying the factors surrounding normal processes of erosion and their acceleration under various types of land use.

It will be necessary to cooperate with the Weather Bureau and to make use of their facilities insofar as possible in analyses of the problems involved on our various project areas. It would seem highly desirable to locate complete climatic stations within watershed project areas to be selected as representative of typical regions without the United States. Let this be considered as a possible development of your phase of the work.

I doubt if it will be possible for us to arrange for part-time employment except as follows: Your appointment may be made effective as of June 1, but you could report for duty on May 15 and work for one-half month in May and one-half month in June. It is highly important that the climatic studies be gotten underway.

The Climatic and Physiographic Research Section was established in July 1935, with Thornthwaite at its head. The Thornthwaite family moved to a house in Arlington, Va., in 1935. There a second daughter, Dorothy, was born on January 20, 1937, and Sally, a third daughter and the last child, was born on October 9, 1939.

Thornthwaite lost no time in trying to locate assistants to help him in his new position as chief of the Climatic and Physiographic Research Section of the Soil Conservation Service. One person he turned to for advice was R. J. (Dick) Russell, a friend from Berkeley days, who was then at Louisiana State University in Baton Rouge. A letter from Russell dated May 19, 1935, stated,

Dear Warren:
Your new position sounds like very good news. I take it that it means final and positive release from Bollinger [from Oklahoma days] and am mighty glad that it has happened. It will be a wonderful thing to have a real research division in the Soil Conservation Service, so power to you.

We have a couple of boys here who might fit into a division of the type you describe. One is excellent, but we don't want to let him go for another year or so. He worked for the Soil Erosion Service last summer, and they wanted to hang on to him.

Thornthwaite immediately began putting together research proposals. He also contacted another friend, this time from the University of Oklahoma, to help him with his new climatic projects. Paul Sears, a botanist, replied in July 1935,

Dear Warren:

So far as the Great Plains trip is concerned, I should like to accompany you, particularly if it comes in September. I see no reason at the present time why I cannot do this since the task of seeing my book thru the press will be well over by that time.

I will be here until the end of the coming week, that is the 27th, after which I can be reached at Norman. With kindest regards and holding myself ready for the privilege of joining you on your plains studies if that seems possible.

On September 20, 1935, the Soil Conservation Service approved Thornthwaite's establishment of a climatic research center in western Oklahoma. The Weather Service was also involved, and two hundred weather stations were established in an area of about eighteen hundred square miles. The fact that Thornthwaite recommended such a dense network of climatic stations is an indication that as early as 1935 he was already realizing the importance of mesoscale climatology. In February 1936, Thornthwaite was informed that he had been allotted $168,000 (a great deal of money in the 1930s) for the project. For this project it was necessary to hire Works Progress Administration (WPA) workers. Evidently, Thornthwaite hired more scientists (nonrelief workers) than he was entitled to, as the following memo (November 1936) from a WPA liaison officer indicates:

Your employment figures for WPA funds as of November 7, show the following: Relief workers: 184 (77%)   Non-relief workers: 54 (23%)

You will note that the proportion of non-relief workers to relief workers in the above statement is considerably in excess of the 10% allowed by FW #SCS-386. Please endeavor to correct this at the earliest possible moment, either by hiring additional relief laborers or by transferring some of the non-relief workers to the payroll for regular funds.

Almost two months have elapsed since the distribution of FW #SCS-386, which states that the 90–10 ratio applies to both old and new WPA funds, and it is incumbent upon the Soil Conservation Service to see that this is strictly adhered to if we are not to be subject to severe criticism by the WPA with possible curtailment of emergency funds in the future.

Very truly yours,
Henry H. Collins, Jr.
WPA Liaison Officer

Another of Thornthwaite's research programs in his Division of Climatic and Physiographic Research involved the Polacca Wash project in Arizona, which was initiated in 1935 and involved physiographic or geomorphological work, rather than climatic work. It was a project of Carl Sauer's and was an outgrowth of his work under Isaiah Bowman's direction through the Land Use Committee of the Science Advisory Board. While Sauer and Bowman were friends and colleagues, Sauer (and Leighly) appeared, at times, to have a poor opinion of Bowman, as the following letter written in 1976 by John Leighly to Geoff Martin indicates:

There was a meeting of the National Research Council in Berkeley in about 1930 or 1931, and in connection with that meeting Bowman was invited to give a public lecture on the campus. The lecture was given in one of the largest lecture halls, and was well attended; I went to the lecture with Mr. and Mrs. Sauer, and we sat together. Bowman's lecture was atrocious: unorganized and full of statements that called attention to himself. When it was over and we were getting up, Mr. Sauer said, "Let's get out of here before someone recognizes us." We did. That was the first time I saw Bowman. (Reproduced with permission of G. J. Martin, Southern Connecticut State University)

The Polacca Wash project ended by being a big headache for Thornthwaite. The chief purpose seemed to be the study of the dynamics of stream erosion by geomorphic methods. Sauer was appointed temporary senior soil conservationist by Thornthwaite in 1935, and that first summer Sauer had three men in the field: a geographer, Dick Normand, and two geologists, Francis Johnson and Perry Reiche. Evidently there was friction among the three men, and Reiche left the project. Johnson carried on for a second field season, mapping the terrain and installing rain gauges, but he appeared to be unhappy with the project and resentful that, as a geologist, he was working under the direction of Thornthwaite, a climatologist. According to papers in the Thornthwaite files, there was friction between Johnson and Thornthwaite. Many letters flew back and forth during the summer of 1936, and Johnson resigned from the project in August, 1936. Thornthwaite stated, in a letter to Bowman dated August 13, 1936, "I am perfectly willing and ready to admit the fact that I do not know all the answers to the question of the relation of geomorphology to the practical problem of soil erosion; for that matter, neither do I know all about climate or ecology."

As a senior geographer, scientist with an international reputation, and president of a prestigious university, Bowman had a good deal of influence. He approved of Thornthwaite's handling of the Johnson affair and did not hesitate to offer him personal advice. Perhaps Thornthwaite learned something from this episode of how to handle a subordinate. It probably also convinced him that geomorphological research was not his forte and that, in future, he should confine his activities to climatology, an area he knew more about.

Although the administration of the Polacca Wash project was far from smooth and some recriminations and harsh words occurred between field researchers and Thornthwaite, the scientific results were of some importance. A monograph entitled

"Climate and Accelerated Erosion in the Arid and Semi-Arid Southwest, with Special Reference to the Polacca Wash Drainage Basin, Arizona," appeared in 1942 that summed up the principal findings concerning the rate of gully erosion and rainfall frequency and intensity in a dry climate.

From the memos quoted earlier and the Johnson episode, it is obvious that Thornthwaite did not enjoy being an administrator in the federal civil service and was much happier doing field research or designing instruments for measuring climatic parameters. He was continually in trouble with his superiors in the civil service. An example is described in a memo dated April 8, 1936, from Lowdermilk:

The objective of this piece of work is splendid. I want to raise a point, however, which should guide us in this sort of work in the future. A project of this nature would better be cleared through the Chief of the Division of Research whereby attention could be called to the other agencies of our Service which might be of assistance. For example, Mr. Fuller should have been consulted in the preparation of your plans. His knowledge of soils generally and of men who are acquainted with soils throughout the country would have been of great assistance in securing the information you desire.

Thornthwaite found that the red tape involved in the Soil Conservation Service also applied to the publishing of his scientific papers. In the Soil Conservation Service files in the National Archives there is a memo to Thornthwaite dated February 15, 1937, that states,

Doctor Bennett's approval of your paper "The Life History of Rainstorms," prior to its publication by the *Geographical Review*, naturally relieves you of your responsibility for having manuscripts approved by the Chief of the Bureau.

Nevertheless, there are several aspects of this matter to which I am constrained to call your attention. The regulations governing the clearance of manuscripts in advance of their publication are prescribed, not by Doctor Bennett or by any other member of this

Bureau, but by the Office of the Secretary of Agriculture. Paragraph 1336 of the Administrative Regulations of the Department states in part as follows:

> " . . . If the material treats in any way of the policies of the or the work of any other bureaus or Departments, it must be submitted by the Chief of the originating Bureau to the director of Information for approval before it is offered for publication. . . . One copy of each article or written address bearing upon the work of the department and prepared for outside publication or delivery should be sent to the Office of Information for reference at least ten days in advance of the date of the publication or delivery. . . ."

Obviously, all members of the staff of this Bureau are obliged to conform in all cases with this regulation. The responsibility for facilitating the handling of manuscripts in accordance with this procedure is mine by virtue of the position I happen to occupy; I am responsible in turn to the Director of Information of the Department.

You will perceive at once that in approaching Doctor Bennett with a request that he approve a manuscript, members of the Bureau's staff impose upon him a responsibility for clearing the manuscript through Departmental channels. He is far too busy, as you know, to attend to such matters—too busy, in fact, to recall that such clearance is required. It will greatly ease his burden, I believe, if you and other members of this staff will cooperate with my office in obtaining advance approval on the manuscripts you desire to present for publication in outside journals.

I am sure you will understand that I have no desire whatever to surround you and other scientists in the Service with hindering rules and regulations. On the other hand, I bear a direct responsibility to the Department's Office of Information in seeing to it that the manuscripts prepared in this Bureau are cleared in advance in accordance with the over-all regulations of the Department.

G. A. Barnes,
In Charge, Section of Information

The following memo included in the National Archives in Washington indicates another of Thornthwaite's problems, the irritating and time-consuming problem of keeping accounts for the Department of Agriculture's finance office:

September 29, 1937

Reference is made to your memorandum dated August 28, 1937, addressed to Mr. F. W. Darnell, relative to our charge of $13.80 for photostat work performed for your Section during the month of July.

We have checked very carefully the photostat work performed for your office during the month of July and find that three jobs were handled. On job orders 3045 and 3052 we find the charges of $3.45 and $2.54 respectively to be absolutely correct. However, on job 3048 you are perfectly justified in questioning the charge of $7.81 as our recheck reveals this charge should be only $1.25. This mistake was evidently made in a mix up of job orders, charging your Section for a portion of work which had been performed for some other Section. We are very sorry this mistake occurred and appreciate that the matter was brought to our attention. We will do everything possible to see that this does not happen again.

In order to correct this error, a credit of $6.56 will be allowed on work performed for your office during the month of September.

Such memos (also in the National Archives) from the finance office were a constant irritant:

Reference is made to a telegram dated September 17, 1937, from you to G. E. Redden, Maysville, North Carolina, which reads, "You may stay on leave until first subject to call." This message is considered personal, per paragraph 58, Government Travel Regulations, and must be paid for by the interested party at the commercial rate of $0.60 plus Federal Tax of $0.03.

It would be appreciated if you would secure a check or money order in the amount of $0.63 made payable to the Treasurer of the United States and forward it to this office in order that reimbursement may be made to the telegraph company in accordance with the provisions of General Regulations No. 40, Supplements 2 and 3.

We shall appreciate your giving this matter your early attention.

Thornthwaite made an attempt to appease the finance officers while devoting most of his time to his research. He continued to correspond with Carl Sauer and inform him of the research project underway in Muskingum, Ohio, and the work of his colleague David Blumenstock. Blumenstock was a protegé of Carl Sauer's who attended the University of California, Berkeley, after graduating from the University of Chicago in 1935. He especially enjoyed the courses in climatology with John Leighly. In 1938, when he was 25 years old and writing his doctoral dissertation on drought, he obtained a job in Washington as research climatologist under Thornthwaite, and many of his later climate writings had their beginnings during the three years he spent in the Soil Conservation Service. One of his most successful books was *The Ocean of Air*, which finally appeared in 1959. When the United States entered World War II in 1941, Blumenstock served as meteorologist and climatologist in various capacities. He was stationed for a while in Hawaii and wrote, with Saul Price, the first account of the climate of the Hawaiian Islands. His very promising career was cut short by illness, and he died in 1963, the same year as Warren Thornthwaite.

In October 1937, Thornthwaite wrote to Sauer,

In our original project outline for climatic studies, submitted to the Inter-Bureau Committee in the summer of 1935, proposed work on climatic change was included. This work was disapproved by the Inter-Bureau Committee with the notation that it was within the province of the Weather Bureau. I have never taken the trouble to make an issue of it, because there remained in our approved plan more than we could do with our staff and funds. The fact that we are definitely prohibited from making such studies would make it impossible for us to use Blumenstock on the job which you outlined. The fact is, in spite of his interest in the tree ring material, I have other work which appears to me as being much more urgent which I would have him do if he were to come on for a time. Our microclimatic study in the Muskingum Valley is in full swing. Corwin has there five hundred complete weather stations, each of which includes a self-recording rain gage in addition to the other instruments such as we had in Oklahoma. Preparation of half-hour interval maps began July 1st, and we have three months of these accumulated so far.

There are thirty or more stream gaging stations in the valley, and we are now attempting to develop better means of relating rainfall and run-off and attempting to determine the influence of cover and land use in run-off. [H.A.] Ireland and Guthe (a Ph.D. from Michigan) are in Ohio, attempting to develop some quantitative basis for determining the run-off characteristics of drainage basins. If Blumenstock were free for the first half of 1938, I would send him first to Ohio, and then have him come in to Washington to work on certain specific assignments in connection with the Ohio work. I would be glad to have your reaction to this proposal, since it appears that we could give Blumenstock a temporary appointment beginning January 1, 1938, or earlier, and terminating not later than June 30th, with the possibility that funds might be found to continue the appointment.

We have all been working on a revision of the Basin Drainage Report published by the National Resources Committee in December 1936. This is in connection with the President's comprehensive message to Congress next January on flood control and water utilization. I have proposed the inauguration of three new micro-climatic studies similar to the one now operating in Ohio. One is in southwestern Iowa; the second, in southeastern Colorado, and the third one in the East Bay section of California. It includes Contra Costa, Alameda, Santa Clara, and Santa Cruz Counties. In a way, a better location for a California study might be the area east and south of Los Angeles, but I thought that other factors made the area selected more attractive.

There is, of course, no assurance that this proposed work will ever get into the revised Basin Drainage Report; and if it did, that funds would ever be appropriated for carrying it out.

Another letter to Carl Sauer, in January 1938, contained the following:

I was pleased to have your letter of December 24th. It is not at all unlikely that nothing more will ever be heard of the proposed climatic project in the Bay region. Just at the present I am expending all of my energy in trying to keep the Ohio project going. Funds will be exhausted on the fifteenth of January, and the extension project which we submitted last September has not yet cleared the various offices. My reason for including Santa Clara County was that I wanted to have a complete watershed in which stream gage records and well records would be available.

The first results from the southern Piedmont are now nearing final form as a bulletin entitled "Principles of Gully Formation." I would like very much for you to examine the manuscript before it leaves our hands. It is quite long, extends through two hundred fifty typed pages, and has many maps and illustrations. If we send a copy to you as soon as it is finished would you have time to examine it within a couple of weeks and return it with your comments? If you feel that you have time to look at it, please send me a wire collect. The idea would be to get your opinion on content and organization rather than on phraseology.

Thornthwaite was busy writing articles and giving speeches during these years with the Soil Conservation Service. A paper entitled "The Significance of Climatic Studies in Agricultural Research" was published in the *Soil Science Society of America Proceedings* in 1937. In this paper, Thornthwaite wrote of his new interest in the relationship of precipitation to evaporation and referred to his earlier paper on climatic classification. He also wrote of his burgeoning interest in the year-to-year variability of climate.

An interesting paper was given in 1937 at Queen's University in Kingston, Ontario, on "Climatic Studies and Canadian-American Relations." Isaiah Bowman was the chair on this occasion. In his speech, Thornthwaite displayed a good knowledge of the Canadian climate. He said that he had "spent several summers in Ontario and determined the size and age of thousands of black spruce trees from Lake Simcoe to James Bay" and indicated that tree rings are a good evaluation of temperature efficiency.

Another paper given by Thornthwaite in 1937 was to the annual meeting of the Association of American Geographers in Ann Arbor. In this paper, Thornthwaite described the "geographical research in the Soil Conservation Service." In it he attempted to show that the work being done in the Soil Conservation Service was of interest to geographers, especially the climatic research being done in Ohio on rainfall-runoff relationships.

During these busy years of organizing his Climatic and Physiographic Division, Thornthwaite kept in touch with John Leighly. In April 1938, Thornthwaite wrote,

> I have delayed in answering your letter of December 26th so long that I believe it is safe to assume that I have gone beyond the apology stage. The arrival of the April *Review* gives me the incentive to write you. Needless to say, I liked your review of Holzman's paper very much, and I must confess, I believe your drawing of the hydrologic cycle is an improvement over ours. As you state in your review, the relation of the air mass cycle to the hydrologic cycle is so apparent and so obvious that there should be no argument about it, whatever. Nevertheless, there is a great deal of argument, particularly it is true that the older hydrologists who have committed themselves repeatedly are reluctant to accept the new interpretation.
>
> Our latest bulletin has just been released by the Government Printing Office. If you have not yet received it, I assume that you will in the course of a few days. It is by C. F. Stewart Sharpe and is called "What is Soil Erosion?" We dislike the title very much, but since it was the choice of the Head of the Section of Information, there was nothing we could do about it. The bulletin is so profusely illustrated that it looks like another bit of propaganda, but I assure you it isn't! I would be glad to have your reaction to it after you have had time to examine it.
>
> By the way, I hope that you have seen Sharpe's new book on *Landslides and Related Phenomena*. Dick Russell had a very enthusiastic review of it in the first issue of *Journal of Geomorphology*. I personally believe just what I said in the *Geographical Review*.
>
> I meant to write to you long ago when the issue of the *Annals of the Association of American Geographers* containing your article on the "Topography of Art" appeared. I wanted to be the first to congratulate you, but as you see, the time passed and I failed to get the letter written. You have probably heard that the paper was the principal topic of discussion at the Christmas meeting in Ann Arbor. The Michigan

geographers were divided into two armed camps, with Dodge for and James against. Hall seemed to be neutral, and McMurry indifferent; all of which is precisely as you might expect. Stanley Dodge told about a letter from his father in which the elder Dodge commented on the break between Sauer and Leighly. I told Stanley that it was amusing that his father should have thought that Sauer had failed to develop during the last thirteen years. Most amusing incident of the conference was when Dick Hartshorne proclaimed that it was time for geographers to declare their independence from "the God who lives beyond the Sierras." Calkins was at the meeting with a well-worn and marked copy of your reprint.

Dave Blumenstock is adjusting himself very well, and is doing very creditable work. I sincerely hope that we will be able to find money enough to keep him on.

Benjamin Holzman, another young colleague of Thornthwaite's, was doing excellent theoretical work on atmospheric water vapor. Using, in part, the Muskingum data, he wrote an article, "Sources of Moisture for Precipitation in the United States," which was published in a United States *Department of Agriculture Technical Bulletin* in 1937. Thornthwaite continued in his letter to Leighly,

The Ohio climatic project is still running and still collecting vast quantities of extremely interesting source material. We are beginning to count the days when Holzman will be back to get to work on the theoretical implications of rainstorm morphology. I myself have been spending my spare time recently in studying rainfall-run-off relations and in developing empirical means of determining soil moisture deficiency and ground water accretions in a seven hundred square mile tributary basin on the Muskingum. The work looks very promising at the present time, although it is revolutionary from the standpoint of classical hydrology.

Thornthwaite published this material in an article entitled "The Hydrologic Cycle Re-examined" in *Soil Conservation* in 1937. Again, in December 1938, Thornthwaite wrote to John Leighly,

On several occasions Blumenstock and I have discussed his future work, particularly his thesis subject and the remaining unfulfilled requirements of the University. He told me this morning that Mr. Sauer approved of my suggestion that Dave [Blumenstock] submit a thesis on drought rather than one on tree rings. Dave reports that Mr. Sauer said that you were pleased at the change, that you never had been particularly happy about the tree ring study. I am writing you to see if you would be willing to discuss the matter with the Dean or Mr. Sauer to see if it might not be possible to save Dave a hundred or so dollars by arranging the meeting somewhere east of the Mississippi.

I am planning to meet Dick Russell in South Carolina next weekend for two or three days' field trip with my men who are working on the erosion problem of the Piedmont. Dick will spend a few days in Washington with us before going on to the meeting in Cambridge. He also plans to stop in Ohio on his way back to Louisiana. Mr. Sauer might be interested in these plans.

During the last several months we have dropped right into the middle of the evaporation problem. Mr. Sauer may have told you about our instrumental installations in Ohio. We are determining the specific humidity gradient in the turbulent layer, and have gotten around the non-linear nature of the gradient by making observations of wind velocity at the two levels. This enables us, as you see, to get a measure of the Austausch coefficient which we find is continually varying. We have an installation at the Bureau of Plant Industry farm just across the river. The final equation is relatively simple, since the Austausch coefficient is expressed in terms of difference in wind velocity at the two levels.

Needless to say, we have made a great deal of use of your paper which came out in *Ecology* a year and a half ago. It was, in fact, the thing that got me started along the present line. As I see it, the measurement of evaporation by determining the gradient through the boundary layer cannot be made successfully until more satisfactory instruments for measuring vapor pressure are available. Even in the turbulent layer, where we have a range of several feet, the instrumental difficulties are very great. In the hope of getting an instrument which will be more satisfactory than the hair hygrometer or the wet and dry bulb I have recently designed a dew-point recorder. A pair of these instruments are now being built, and it is barely possible that they will give us better figures on the moisture concentration in the atmosphere.

We have recently prepared a paper summarizing the climatic research in the Soil Conservation Service. It is to be published shortly in the *Monthly Weather Review*. There is still a little time, however, for making changes in the manuscript, and I am sending a copy of it to you in the hope that you will find time to read it and may be able to make comments which will improve it.

As can be seen by the above correspondence, Thornthwaite was becoming increasingly interested in the loss of water from land surfaces to the air by evaporation and transpiration. In this work, he was being aided by David Blumenstock and Benjamin Holzman. In 1938, Thornthwaite and his two young colleagues, Holzman and Blumenstock, wrote in the *Monthly Weather Review* "Climatic Research in the Soil Conservation Service." In this paper, the microclimatic data from the five hundred weather stations in the Muskingum valley were analyzed to show maps of, for example, twelve-hour rainfall on the whole of the Muskingum drainage basin. The paper includes an interesting discussion of the importance of evaporation in the hydrologic cycle, a concept not universally acknowledged.

In the articles Thornthwaite wrote at the time are numerous references in German, which he was able to read in the original language, thanks to his training with Carl Sauer. It is also apparent in these articles that the study of climate for itself, and not only for its relationship to soil erosion, was becoming important to Thornthwaite.

During 1938–1939, Thornthwaite was completing his yearlong study of evapotranspiration from a grass-covered area near Arlington, Virginia, and was writing the results for publication. The study had been a great success, and the results were far better than any that had been obtained previously. In this study, Thornthwaite introduced a new method of obtaining field observations of evapotranspiration that allowed researchers to get away from the water-filled evaporation pan or soil-filled, vegetation-covered pot experiments that had been in use for more than two hundred years.

Thornthwaite, himself, was enthusiastic about his progress in developing instruments sensitive enough to provide profile observations of wind and atmospheric moisture for a fairly continuous period of time. He sent a copy of the manuscript he was preparing to John Leighly. Leighly was greatly impressed by this contribution and wrote back almost immediately the following words of praise:

January 12, 1939

Dear Warren:

I've been so consumed by enthusiasm since reading the manuscript on "The determination of evaporation from land surfaces" that I have not been able to rest until I could write you and express some of that enthusiasm. I've read a good deal of literature on evaporation, but nothing that has been so enlightening as this. Consider that I throw my figurative hat in the air and emit a figurative hooray. The problem seems to be solved, and your further improvements will be only improvements in instrumentation. I'm already seeing in my imagination such installations as yours over corn fields and irrigated alfalfa fields and above the crowns of forests; and a new class of climatic data from everywhere, namely evaporation from the natural surface of the ground. It's a magnificent prospect.

George Carter of the Johns Hopkins University recalls that, about this time, Thornthwaite had an undergraduate student, Maury Halstead, making observations for him in California. Instruments were set up to measure the moisture flow in the Berkeley Hills. Thornthwaite wanted to get to Berkeley to check up on the observations but was prevented from doing so by some governmental regulation, and so he simply took a pick-up truck, drove across the continent, and turned up in the geography department at Berkeley.

Thornthwaite continued to use Leighly as a sounding board

and critic of his ideas and writings. In 1941, Thornthwaite was busy preparing a substantial article on "New Duties for Climatology," which actually never appeared as such, though portions of it did appear in later writings. He kept sending drafts of the article to Leighly for criticism and review. Because Leighly was in California and Thornthwaite was in Washington and it took several days for the mail to go coast to coast, it often happened that Thornthwaite would mail his revisions of his manuscript before Leighly could complete editing the earlier version. This practice finally frustrated Leighly enough for him to write the following on November 23, 1941:

Dear Warren:

This patched page is a monument to impatience. I wrote a letter to you yesterday in which I spoke out rather sharply about your sending me draft after draft of your "prospectus," each one arriving just about at the time when I had finished going through the previous one, and was ready to write you some criticisms and suggestions. The arrival of the third draft yesterday was too much. So I have cut off the first paragraph of yesterday's letter, and am replacing it with this.

Seriously, however, I have no more time to work on your manuscript. I have spent a good share of my time the past two weeks or so on it, and most of that time was wasted, since you had new drafts before I could get anything to you. Now I have a lot of things to do before I leave in less than a month, the most pressing of which is a paper for Pete Burrill. Abstracts of papers should be in Jimmy James's hands tomorrow, and I haven't even an outline of what I shall present. I also have chores to be done about the house, that I haven't had time to catch up with during the semester.

For whatever it may be worth to you now, here is an outline I made on the basis of your second draft, my copy of which is enclosed with many remarks on wording and other matters. Enough of that draft is preserved in your latest one to make these remarks of some possible use. My principal objection to the second draft was that it was too shapeless. Hence the following outline, which rearranges the material in part. Pages refer to the handwritten pagination on the copy enclosed.

The letter contains another page and a half of suggestions, plus the manuscript with numerous corrections and suggestions. Although the letter's tone might seem critical of Thornthwaite, he took it good-naturedly; very shortly, he suggested that a portion of that paper should be submitted under joint authorship with Leighly as "Status and Prospects of Climatology." It was published in 1943 in *Scientific Monthly*. This landmark paper discusses some of the history of climatology in the United States, the importance of work in microclimatology (actually an early recognition of the importance of topoclimatology), and ends with a strong section outlining the structure and operation of an institute for climatic research, which became almost a blueprint for Thornthwaite's Laboratory of Climatology.

Possibly one reason for the delay of, and finally the decision not to publish, the "New Duties" paper was Thornthwaite's rather severe criticism of another government agency, the U.S. Weather Bureau. He discussed the fact that meteorology was really only climatology up to the time of the development of the telegraph in the late 1840s. Then climatology entered into a long period of decline as all the energies of the meteorological community were directed toward forecasting. Only agricultural climatology, championed by dynamic individuals such as Cleveland Abbé, was able to develop in the latter part of the nineteenth century. The following paragraphs are quoted from Thornthwaite's unpublished manuscript on "New Duties for Climatology":

Daily weather forecasting is in the process of being improved but climatological activities in the United States are at about the same level today as they were 50 years ago. Furthermore, there is no possibility of immediate improvement because there is no source of trained climatologists. The trained meteorologists are not prepared to deal with problems in physical and applied climatology. Furthermore, they are more or less indifferent to these problems, and are often scornful of them. It is hopeless to expect the Weather Bureau or the meteorologists in the country to take the lead in a revitalization of climatology. . . .

The failure of the Weather Bureau to exercise leadership in climatology is due largely to the fact that it did not recognize that the materials and methods of dynamic meteorology and physical climatology are of necessity different, and that each field must be cultivated by men especially trained for the task.

The geographers, who have maintained a strong interest in the field of climatology, have studied climate almost exclusively as a key to the regional differentiation of the earth, and consequently have concentrated on descriptive climatology and on the classification of climate. The study of descriptive climatology has been diligently pursued and bulletins and books on the climate of particular regions and places are countless. The many textbooks on climatology in English are largely descriptive and have more or less similar organization, giving information on temperature, rainfall, humidity, pressure, and wind for various parts of the earth. University courses in climatology in this country pretty generally follow the textbook outline. At the same time there have gradually emerged, without the climatologists being fully aware of the fact, new fields of climatological endeavor and new duties for climatology.

There is urgent need for trained climatologists to work on applied problems. The training should include a thorough grounding in mathematics, physics, physical meteorology, and statistics, as well as descriptive climatology and some geography. It should cover all that is known of physical climatology. It should also include sufficient training in a special field, such as biology, engineering, conservation, social science, or medicine, to develop competence in dealing with various applied problems.

For many of the problems in applied climatology, solutions are not now possible but must await the completion of some phase of the research in physical climatology. Approximate solutions of some applied problems can be obtained immediately but, in general, little success with the applied problems is possible until much additional fundamental research has been completed.

The time is ripe for a renaissance in climatology. Although the Weather Bureau is the logical agency to exercise leadership in this field, under the present limitations of outlook, and staff, and with the service routine under which it labors there is no prospect for a renaissance to get under way there. The leadership cannot be provided by any Government Bureau but must come from outside. It was the modest support of the Guggenheim Fund that gave the initial boost to meteorology. A similar boost is now needed for climatology. (Unpublished manuscript in Thornthwaite files)

The United States entered the war at the end of 1941 with the Japanese bombing of Pearl Harbor, and most of the members of Thornthwaite's division went into the armed forces. Thornthwaite was a pacifist by nature and had no desire to be part of the war. For most of the war years, he was a rather lonely figure in the South Agriculture Building in Washington.

Fred Kniffen recalls a strange incident that happened to Warren in 1941: "It seems he was accused by the Secret Service of patronizing a music store in Washington operated by and the center for a group of subversive activists. I testified to the fact that Warren was innocent of anything but a great love of good music expended in buying records from a convenient and well-stocked store. He got off! I do think he was capable of a 'subversive' thought or two, but never took action" (personal communication with Mather, 1990).

Many years later, Thornthwaite needed a security clearance for some work with the government, and it came to light that he had been member of the Washington Bookshop Association, an organization that had been cited as Communist and subversive by the U.S. Attorney General. Thornthwaite explained his actions to the Appeal Division of the Eastern Industrial Personnel Security Board as follows:

While I was Chief of the Division of Climatic and Physiographic Research in the Soil Conservation Service, in Washington, my secretary . . . reported to me that there was a book shop on Connecticut Avenue where books and phonograph records could be obtained at a substantial discount. I visited the Book Shop on March 22, 1941 (according to a cancelled check stub in my possession), for the purpose of seeing what record albums were available. I located four albums of Gilbert and Sullivan operas which I wanted very much. The retail price on these records amounted to something like $55.00, which was beyond my means. But the clerk explained that if I cared to join a book club I would qualify for a substantial discount. The saving was substantial because the check that I issued on that day to the Book Shop is for only $29.40. A part of that payment was probably the membership fee and the remainder was payment for the albums. I don't remember signing anything but I suppose that I did and by so doing acquired the membership. I should emphasize that my check was made out to the Book Shop and not to the Washington Bookshop Association. Since I could not afford to gratify my taste for books and records further, I never was in the Book Shop again. I took it to be just what it represented itself to be; a book club to save its members money, like the Literary Guild or the Book-of-the-Month Club, to both of which I have had memberships at one time or another.

When we first went to Washington, I became a member of a buyer's club which enabled us to obtain many things at various stores at very large discount. Through that club (whose name I do not remember) we were able to save several hundred dollars on our new furniture and rugs. As far as anyone could tell, this Book Shop was the same kind of club. It looked just like any other book store, and I am sure that no customer who walked into it from the street as I did could have recognized it as Communist and subversive. I didn't know anything about Communism or subversion in those days, and I don't know even now what the Washington Bookshop Association was up to or why it was cited by the Attorney General. The clerk who waited on me gave no hint that the store had any purpose other than to sell books. I saw no one else in the store. And, there is absolutely nothing more to my membership in the Book Shop than this; I made a chance purchase there on March 22, 1941, on which I saved more than $25.00 by joining their club. I was never in the store again either before or after that day and I never had any occasion to know anything more about it than that.

A formal hearing on the charge took place on August 10, 1954, and the Appeal Division of the Eastern Industrial Personnel Security Board decided that granting a clearance to view classified information would not compromise the security of the United States. Thornthwaite had been vindicated, but the episode left a scar. He was hurt that some question had been raised about his loyalty. More significantly, he was greatly disturbed by the whole process of "guilt by association" and the feeling that one was guilty until proven innocent.

His daughter Elizabeth recalls that Warren took leave from the Soil Conservation Service in 1941 to return to Michigan for a brief time, though she was unaware of any significant reason for this short-term change of location. Denzil and the two youngest daughters went to live with her parents, the Slentzes, in Clare, Mich., and Elizabeth lived with an uncle in Muskegon, Mich. The family remained in Michigan only for the winter of 1941–1942, then moved back to the Washington area—to Hyattsville, Md., and then shortly to the house in College Park, which they kept for the remainder of their lives. Thornthwaite again took up his job in the Soil Conservation Service.

The 1941 Yearbook of Agriculture—the sixth yearbook of the Department of Agriculture—entitled *Climate and Man*, was a monumental treatise on the subject. It contained five sections: climate as a world influence, climate and agricultural settlement, climate and the farmer, the scientific approach to weather and climate, and climatic data. It is obvious that Thornthwaite, as well as his friends Leighly and Sauer, had a great deal to do with the content of the book. The chapter entitled "Climate and the World Pattern" was coauthored by Thornthwaite and Blumenstock. Thornthwaite also wrote a chapter on climate and settlement in the Great Plains, Sauer wrote a chapter on "Climate of the Humid East," and Leighly a chapter on "Settlement and Cultivation in the Summer-Dry Climates."

Thornthwaite's collaboration with Leighly was so successful in those days that there was serious consideration of a book to be entitled "Physical and Applied Climatology" to be written by Thornthwaite and Leighly (with a section on statistics in climatology to be prepared by Blumenstock). A detailed outline was prepared and submitted to the McGraw-Hill publishing company, and a favorable response was received from the publisher. Clearly, no similar book was available in the United States at that time, and none appeared for nearly thirty years. This indicates how far Thornthwaite was ahead of his colleagues in recognizing what was needed if climatology was to take its place as a rigorous physical science rather than as merely a descriptive addition to geography or meteorology. It is a shame that World War II, bringing the need to undertake other war-related activities, prevented further work on the idea of a book on physical and applied climatology. Had such a book been available at that time, it might have led to a rapid advance in the science of climatology, an advance that ultimately required more than three decades to achieve. The outline of the book, preserved in the Thornthwaite files, included a section entitled "Physical Climatology" that would introduce the history of climatology, including the great names, such as von Humboldt, Maury, and Köppen. There would be chapters on temperature, on the water cycle, precipitation, drought, fluctuations in climate, and microclimate. A second section would outline the methods of analysis of climatic data, including frequency distributions, time series, and the analysis of areal distributions, topics that certainly would have been unique at the time. A third part was to be called "Statistical Evaluation of the Relationship of Climate to Biological and Other Problems." This would include climate and vegetation, soils, floods, irrigation, and even medical climatology.

Although the publisher was very interested in the outline and urged the authors to proceed with the writing of the book, it never materialized, probably because both Thornthwaite and Leighly were too busy doing other things. Unfortunately, shortly after this, there was a cooling of the Thornthwaite-Leighly friendship, possibly, in part, a result of their joint paper, "Status and Prospects of Climatology," which was eventually published in *Scientific Monthly* in 1943. This paper had its origin during Leighly's time in the Soil Conservation Service, but when John moved to Grand Rapids, Mich., to teach at the Air Force Weather School, Thornthwaite produced a major revision of the paper (at the editor's insistence) without consulting or advising Leighly, who later objected to some of the changes.

In the Sauer files in Berkeley, there is some interesting correspondence between Sauer and Bowman in which Thornthwaite's name appears concerning the position of chair of the Geography Department at the Johns Hopkins University. Evidently, Bowman offered Sauer that position, and Sauer replied cordially although he declined. He did suggest others, however (May 21, 1944):

> I still think that Thornthwaite might do very well with yourself as elder prophet. The Washington post was not altogether good for his proper growth, but he is excellent on climate (when he doesn't go back to do battle too strongly for his system of classification), he is a good demographer and a person who can think intelligently by means of maps. I thought of him at once when I read your paragraph on biostatistics. He can work prodigiously and is happier in that than in the administrative duties which he has had and to which he is not well suited.

It is probable that Thornthwaite never heard of this possible job offer. Bowman replied on 27 May 1944 (below), indicating his reservations about Thornthwaite. It is unclear what Bowman meant by "rigidity of mind" and his comment "queer way of talking" may refer to Thornthwaite's convoluted sense of humor! He wrote: "It may be that Thornthwaite will do for a place here, but my present feeling is against it. He has a certain rigidity of mind and a queer way of talking that leaves his listeners (not I alone) somewhat mystified as to his meaning."

While Thornthwaite was working in the Soil Conservation Service during the period 1939–1945, there was little contact with researchers in Europe, especially in Germany, because of the war. For that reason, he was unaware that German climatologist Rudolf Geiger was also researching the properties of the climate of the layer of air near the ground (in German, *das Klima der bodennahen Luftschicht*). Because Rudolf Geiger's name is well known in the field of climatology, the story of his life and work must be mentioned.

Geiger belonged to the same generation as Thornthwaite, and there were many parallels in their lives. Both were impressed by the climatologist Köppen, and for both climatology was the great love of their lives. A later chapter will explain how the two climatologists became good friends. Geiger was a young colleague of Köppen's and collaborated with him in much of Köppen's later work. The following material is from a letter from Geiger to one of the authors in 1973:

Rudolf Oskar Robert Williams Geiger was born August 24, 1894. [About his many names, he wrote, "By the way, in 1950, five years after the war, when I came to the United States, every authority changed the "Williams" in my name to "Wilhelm" even in official documents and my sincere assertions that I really had the name Williams were in vain!"] My father was a linguist and my three additional names were given to me in honour of his best friends at different universities. So I was called Williams after Prof. A. V. Williams Jackson, professor of Indo-Iranian languages at Columbia University in New York. I was the youngest of five children. My elder brother was Hans Geiger, the famous inventor of the Geiger-counter. I was educated at the "Humanistischem Gymnasium" in Erlangen (my birth place), from which I graduated in 1912. During 1912–1914, I studied mathematics and physics at the universities of Erlangen and Kiel. During 1915–1918 I was a soldier in the army as an artillery man. After the war I again studied at Erlangen and earned the Ph.D. degree in 1920. In May 28, 1919, I married Irmgard Klippel, the daughter of the first mayor of Erlangen. We had five children (son, daughter, son, daughter, son), and lost the middle son during the second world war. And during the years 1921–1923 I was an assistant at the "Physikalisches Institut der Technischen Hochschule" in Darmstadt. From 1923 to 1934 I was a meteorologist at the "Bayerische Landeswetterwarte" in Munich. In 1927 at the University of Munich I published the first edition of the "Klima der bodennahen Luftschicht." In 1930 I was a member of an expedition to the West-African coast studying the upper air conditions west of the Sahara for the flights of the Zeppelins to South America. In 1934–1937 I was "Observator" at the Meteorological Institute of the Forest "Forschungsanstalt" München and during 1937–1945 Professor at the "Forstliche Hochschule" Eberswalde. (This period of course included the years of the second world war 1939–1945).

In 1945, while my family was fleeing from the Russians, I became a prisoner of war in the American sector of Germany. In 1946, I became a teacher of mathematics at a ladies' school at Erlangen. During 1947–1948, I was a meteorologist in the weather service in the American sector of Germany at Bad Kissingen and in 1948 a professor of Meteorology and director of the Meteorological Institute of the university. I am still working with great pleasure in my beloved climatology!

Although he did not mention it in that letter, Geiger told one of the authors of this book that Thornthwaite had literally saved his life in the difficult postwar years. The two corresponded after the ending of hostilities, and Thornthwaite, realizing the extreme shortage of food in Germany, sent parcels of food to the Geiger family. Geiger also told Thornthwaite about the loss of his son during the war.

Geiger's contribution to the science of climatology and especially microclimatology is remarkable. His book *The Climate Near the Ground* is a classic and mandatory reading for all climatology students. It went through several editions and was translated into English in 1950. Rudolf Geiger died in 1974, and his place in the history of climatology is assured. He never lost his great enjoyment for his "beloved climatology."

Many of Thornthwaite's publications during this time were in the *Transactions of the American Geophysical Union* rather than in geographical journals. This was indicative of the fact that he was becoming more quantitative in his approach to climatic problems and was learning the value of mathematics and science from his younger colleagues. Thornthwaite also continued his interest in the classification of climates, and in 1943 he wrote an article for the *Geographical Review* on "Problems in the Classification of Climates."

David Miller, a well-known climatologist and later professor at the University of Wisconsin–Milwaukee, knew Thornthwaite during his last years in the Department of Agriculture and recalls,

> Some time after I was transferred to the Quartermaster General's climatic branch in Washington and probably in 1944, Paul Siple and I visited Warren in his office in USDA. Our branch was involved in working out heat balances for soldier clothing, and developing guidelines for field commanders to specific items of uniforms, both hot and cold; it was affiliated with other extreme-climate people— high-altitude, etc.—in OQMG [Office of the Quartermaster General] and drew on mountaineers and explorers like Siple, who had at the time lived longer in the Antarctic than anyone else. I don't recall in detail, but suppose we consulted with Warren on mapping climatic elements involved in clothing issue.
>
> We were, of course, well aware of Warren's work with turbulent heat exchange, and the research on the turbulent transfer of latent heat demonstrated in his long measurement project on land where the Pentagon then stood. This grew out of his work in the SCS [Soil Conservation Service], before that agency fell largely into the hands of engineers who took a structural approach to erosion and problems. Warren's group did a great deal of fine solid work in hydrology, terrain effects, and meteorology; e.g., the first study of thunderstorm rainfall, with synchronized rain gages on high time resolution. I suspect that both his interest in mechanical turbulent exchange and in the great drought of the mid-30s were contributors to an interest in evapotranspiration. Something in this line had led to his collaboration with people in Mexico, which then had been going on several years, and which I believe provided him with measurements of ET [evapotranspiration] on a field or project scale. These were

better than Warren could have gotten from American irrigators, operating with water rights that required them to use it or lose it regardless of crop needs. Plains ET also entered in his finding that foresting the Plains would not at all affect the rainfall. His work with Holzman showed that most vapor went into continental air masses and right off the continent.

Several of his former staff, like Holzman and many others, had joined the military, and so Warren worked mostly by himself in a large, imposing office in the old South Agriculture building. I suppose that this period gave him the chance to work out his seminal paper of 1948, three or four years later than the visit I am remembering.

Thornthwaite had been working on a revised classification of climate for a number of years, as could be seen in his published articles during the 1940s. The central concept of the new revision was that of potential evapotranspiration, a new term in climatology. He recognized that two aspects of evapotranspiration had to be identified. Actual evapotranspiration is the water that evaporates and transpires under variable soil moisture conditions while potential represents the water that would be lost by evaporation and transpiration if an unlimited supply of water existed. He summarized his work on evaporation and transpiration by explaining how the amount of water that leaves the surface of the earth differs from that received by the surface of the earth in precipitation. Precipitation is a physical process and is relatively easy to measure. Evapotranspiration is also a physical process—the flow of water from earth to atmosphere—but it must be studied by biological methods unfamiliar to the meteorologist or climatologist.

Thornthwaite analyzed the data he had gathered on the use of water by irrigated crops in different parts of the country. His analysis of the available data allowed him to develop a formula to compute the potential evapotranspiration of a place from information on its temperature and latitude. The resulting formula that Thornthwaite supplied lacked mathematical elegance, as he himself admitted, but was easy to use with the aid of a nomogram that he devised and supplied in a reduced form in the article he published describing this work.

In a series of articles in the *Transactions of the American Geophysical Union* beginning in 1944, Thornthwaite showed that the monthly water balance of a place could be described by the relationship between the need for water, or potential evapotranspiration, and the supply, or precipitation, as subsequent generations of geographers are well aware.

It was this concept of potential evapotranspiration and the water balance that made his article "An Approach Toward a Rational Classification of Climate" so important. Published in the January 1948 issue of *Geographical Review*, it was without doubt the most influential paper that Thornthwaite published. It set forth a new approach to climatic classification, but perhaps more importantly it outlined the water balance bookkeeping procedure for evaluating the hydroclimatic factors of water surplus and deficit, water storage in the soil, and actual evapotranspiration.

Although only forty pages in length, the article was brilliant in its presentation of Thornthwaite's concept of the water balance or hydrologic cycle as applied to the United States. After outlining the role of potential evapotranspiration as a climate factor, based on his earlier research on the topic, Thornthwaite illustrated his water budget model with a table in which potential evapotranspiration (PE) or water need was compared with precipitation or water supply. In this model the amount of water capable of being stored in the soil was assumed to be 10 centimeters, although he stated that a different soil moisture storage could be used if considered appropriate. A bookkeeping procedure permitted the estimation of monthly actual evapotranspiration, water deficiency, and water surplus. Thornthwaite did the water budget calculations for all the climatic stations in the United States manually, using a nomogram (graph) to simplify the estimation of monthly potential evapotranspiration.

Because the original aim of the paper was a climatic classification, Thornthwaite used PE, deficiency, and surplus values to derive a moisture index for that location. Climatic types (perhumid to arid) were based on the moisture indices and, as in his 1931 classification, were given letter names (A, B, C, etc). A temperature efficiency index was also developed using the PE value. Climatic regions based on thermal efficiency were also given letter names: (A = megathermal, B = mesothermal, etc). The classification was cumbersome and, except within the geographic community, was little used in the world of science. The water budget model, on the other hand, proved to be of great value to scientists in many fields and continues to be used today, more than fifty years after its publication.

Although the war prevented Thornthwaite from knowing it at the time, other scientists—in particular, Howard Penman in England—were also working on the concept of potential evapotranspiration. Ken Hare, who later became a good friend of Thornthwaite's, recalls,

> During the war I spent a while (in 1943 to 1945) working on soil trafficability, and became familiar with surface energy and moisture balances. I worked at Rothamsted with Howard Penman who developed soil moisture budgeting as an operational tool. I soon learned that Penman, being a physicist, had no use for Warren's loose approach to experimentation. Penman also thought of plants and vegetation as passive mechanisms, but nevertheless Penman's budgeting assumptions were amazingly like Thornthwaite's. Both had the idea of potential evapotranspiration, and both assumed that soil at field capacity had 10 cm of available water, and that actual evapotranspiration equalled potential until that water had largely gone, regardless of plant cover.
>
> I came to Canada in late 1945 to McGill University in Montreal. In 1948, the Geographical Review published Warren's potential evapotranspiration paper. I was thunderstruck. I was sure it was

basically right, but the derivation of the celebrated formula was obscure—in fact, not given (it never was, as I recall). So I went down to Seabrook, where Warren then had his laboratory. At our first meeting, after the opening civilities, I said cautiously (I was only 29): "Dr. Thornthwaite, I have some reservations about the way you handled the water balance in your recent paper." And he replied (slowly, looking over his half-spectacles): "I'll bet you haven't got half as many reservations as I have!"

Thornthwaite could be abrupt to those who questioned his climatic water balance and the concept of potential evapotranspiration. In fact, he did not take criticism easily, and if he felt that it was not well thought out, would often reject it out of hand. However, he was willing to listen to and even debate a point if he felt that the critic could offer valid advice and help. This point is well illustrated by some interesting correspondence carefully saved in Thornthwaite's personal files between Homer Shantz (a noted plant and soil scientist) and Thornthwaite in the late 1940s concerning the development of his concept of potential evapotranspiration and the water use of plants. Shantz was nearly a generation older than Thornthwaite, but by that time they had been good friends for well over a decade.

Shantz asked Thornthwaite about the effect of land use on precipitation because he was interested in the use of fire in the management of the brush ranges in California. Shantz wondered if the dryness of the climate would affect the answer or if surface soil temperature would affect the resulting precipitation amount.

In his answer, dated July 8, 1946, Thornthwaite wrote,

> As long as the root zone of the soil is well supplied with water, the amount of water transpired from a completely covered area will depend more upon the amount of solar energy received by the surface than upon the kinds of plants.
> I believe that the water needs of a meadow are the same as of the

forest that was cut down to make way for it. To the extent that tree roots go deeper than grass roots, trees will transpire for a longer time during drought before the moisture in the root zone is exhausted.

There is no real evidence anywhere that precipitation has been changed in the slightest by anything man has done to the vegetation.

In his answer of July 16, 1946, Shantz praised Thornthwaite for the manuscript entitled "Climate and Moisture Conservation" to be published the following year in the Annals of the Association of American Geographers. Then, commenting on Thornthwaite's statement that plant water need depended more on solar radiation than on the kind of plant, he wrote,

It depends on both. For example a field of millet would use ⅓ as much water as a field of alfalfa under similar and comparable conditions. Light would seldom vary as much from place to place as that. However a meadow may use water more rapidly than a forest. Trees in Eastern Colorado continued green to August while the short grass often dried up by July 1 or 4. On the other hand there is a general impression that grasses use less water than they do.

I still think you would be on solid ground if you took into consideration that plants also vary in use of water from place to place as does sunlight! . . . We have a lot of crops which require a drought to produce a good crop. Wheat is best when it just gets through and has an atmospheric drought—and very nearly a soil drought to ripen. In other words—heavier and better wheat crops are grown on "dry farms" than on irrigated farms! By that I do not mean that your efforts to stop drought are not important.

Thornthwaite does not seem to have answered this letter until April 10, 1947, when he wrote,

Thank you very much for having taken the time to read the manuscript and for having given highly pertinent suggestions. You will be interested to learn that a critic in B. P. I. [Bureau of Plant Industries] criticized so severely my unorthodox treatment of drought and the water problem that the editor of the *Agriculture Yearbook* was afraid to publish it. Doesn't that sound familiar? It has been worked over again somewhat and will appear in the June *Annals of the*

Association of American Geographers. I hope it meets with your approval. At the same time I have written a long article for *Geographical Review* which will probably come out in October.

In your letter you say, "A field of millet would use 1/3 as much water as a field of alfalfa under similar and comparable conditions." I assume you are generalizing from your pot experiments on moisture requirements, and I am inclined to doubt if there is any such difference under actual field conditions. Otherwise, how does it happen that during a dry spell the soil becomes equally dry under different types of cover?

Shantz answered that letter on May 24, 1947:

I have thought a lot about our inability to understand each other on this "use of water by vegetation matter." You say, "I assume you are generalizing from your pot experiments . . . I am inclined to doubt if there is any such difference under actual field conditions." I'm not the only one who has generalized! And I fear there are others which have been adopted for these computations! I fear the "kettle and the pot" are equally black!

" . . . how does it happen that during a dry spell the soil becomes equally dry under different types of cover." Of course that is true. But why can they grow sorghum at Akron [Colorado] and cannot grow alfalfa? Alfalfa uses up the water rapidly—sorghum or millet does not. You see it is rate of use. Where the water amount is equal one crop will grow for its full season and the other goes out in a few weeks.

Surely it's unsafe to assume that the pot experiments do not come reasonably close to the field results. We have checked that many times. The physiology of the plant is always an important matter in the production of any crop.

On June 28, 1947, Thornthwaite wrote back acknowledging the fact that he may have upset Shantz by what he had said in his previous letter, but still arguing his point:

I would give a lot to be able to talk with you at length about our notions of water use of vegetation. I feel confident that we do not differ fundamentally in our ideas. I am sure from your letter that I did not make myself clear in my previous letter. I realize that they do grow sorghum at Akron and cannot grow alfalfa, and I have puzzled a great deal about the reason. Don't you think that it is because the period to actual maturity of sorghum is very much shorter than that of alfalfa?

I am extremely sorry if there was anything in my previous letter that caused offense. I certainly had no intention of offending you.

Shantz explained his objections in a letter to Thornthwaite marked "personal" on May 30, 1948:

> I'm in a quandary regarding your paper, "An Approach Toward a Rational Classification of Climate." It's a lot a fine work—and I'm sorry I cannot approve of your method of arriving at the base on which the superstructure is placed. Perhaps it would be well to go over some of the points which impress me as wrong.
>
> [Page] 57. "The only method so far developed . . ." This brushes aside all the work of Cole and others on the High Plains and also everything based on soil moisture determination made by Briggs and Shantz and by dozens of other investigators. With no adequate reference or consideration you list a few methods and then lump them together with the comment, "These methods are highly artificial, and generalizations."
>
> [Page] 60. "Transpiration and growth are both affected in the same way by variations in soil moisture. Both increase with increase of available water in the root zone of the soil to an optimum." This is an entirely false statement. Under field conditions water content varies from field capacity to the wilting point. Between those two extremes there is no difference in transpiration rate or in growth rate. However growth and transpiration are not closely correlated. There is no increase to an optimum. . . .
>
> Now the general principle of the paper is fine and if it had been based on a reasonable physiological basis the results would have probably closely approximated the facts. I'm not surprised that it was not taken by agriculture—but I think that Geographers will be misled for a long period of time since they will not check the Plant Physiology. . . . I hope you will not be too mad at me for being, from my point of view, perfectly honest with you in this matter. . . .
>
> The thing that worries me the most is that you have so much faith in the generalizations that you conclude that a dry short grass area

loses as much water to the air as does a forest of beech—if I have not misread some of your statements. . . .

I would not send this to you if you had not asked and I hope it will have no effect upon our long friendship. I admire the general paper and hope you can find a better base for your basic formula.

After another comment that this letter was personal, Shantz concluded with a brief P.S.: "If this letter is too strong for you, send it back and I'll try to be more *moderate*!"

Thornthwaite did not answer the May 30 letter until September 17. He then wrote, "Because of a completely disorganized household [both of his wife's parents died within a month of each other during the summer], I have just now seen your two letters of May 30 and June 10. I can't do more now than to explain why and to say that I am not mad; I admit your rightness in some of your criticism, but I disagree violently with you in some others. On some points I think that you misunderstood. Later on I hope to spend some time on your letter—who knows, maybe I'll be in the west this winter and might see you."

This was indeed a welcomed letter as far as Shantz was concerned, and he wrote back on October 12, "I was happy to have your letter, for I feared I had drawn too heavily on your good nature—and that would have made me very unhappy. . . . Don't spend too much valuable time trying to convert me. If we can have a few hours together some day we can do more than by a lot of correspondence."

The correspondence went on for another nine years at a rate of about one letter a year from each, but none of them ever touched again on the 1948 paper or the differences between them on the water use of vegetation. It became more firmly fixed in Thornthwaite's mind that energy from the sun was the prime cause of evapotranspiration and, as long as water was readily available to the plants, the type of vegetation cover would be unimportant. Whether Shantz ever understood and accepted the real difference between potential and actual evapotranspiration cannot be determined from the remaining correspondence, which dealt mainly with the increasing infirmities of age, trips taken, and children and grandchildren. However, in 1948 Thornthwaite had other problems on his mind; he had made a major career change and had left the federal government.

Chapter 4

The Move to Seabrook and an "Institute for Climatic Research"

In the early 1940s, Thornthwaite became increasingly interested in the changes developing in the fields of meteorology and climatology. A 1943 article he coauthored with Leighly entitled "Status and Prospects of Climatology" stated that although meteorology had originally been concerned with all phenomena between the earth and the heavenly spheres, the growing importance of aeronautical meteorology was restricting the definition of the field to the mere prediction of the variable changes in weather from day to day. The article said,

Through its exclusive association with weather, meteorology has lost an important part of its former content. This lost content has to do with the regularly recurrent or periodic changes in the physical state of the atmosphere, from day to night and from summer to winter. The seasonal rise of temperature in spring is a climatic phenomenon, recurring every spring; its average rate, and the temperature to be expected at any particular date, are elements of the local climate. The conditions of the atmosphere expected at any place on the basis of experience constitute its climate. It is climate, therefore, that has been lost from view as it [meteorology] assumed its present state. (Reprinted with permission from *The Scientific Monthly* 57: 457, © 1943 American Association for the Advancement of Science)

Before the invention and spread of the telegraph, meteorology was primarily climatology, for it was not possible to attempt a synoptic view of weather phenomena or to use such an approach to formulate a weather forecast. However, shortly after the availability of the telegraph, simultaneous weather observations from a number of places made it possible to identify the location, movement, and characteristics of individual storms on successive days. The public thus became more interested in the production of weather forecasts, and during the 1870s, many of the leading countries of the world established official meteorologic services to satisfy that growing need. According to Thornthwaite and Leighly (1943), "The attention of meteorologists was shifted; whereas in the 1820s the object of investigation by the science of meteorology was 'climate,' by the 1870s it had become 'weather'" (p. 458).

One of the initial charges given to the U.S. Weather Bureau at its inception was to make those meteorological observations necessary to explain climatic conditions across the United States. To meet this requirement, the Weather Bureau felt the need to develop rigid standardization of instrumentation and of the manner in which instruments were to be exposed. From a cost viewpoint, only a reasonable number of observing stations could be established. Thus, the Weather Bureau staff decided to eschew the measurement of local or microclimates and to obtain fairly standardized observations over uniform surfaces uninfluenced by the diverse mosaic of landcovers that exist in nature. The climates of the valley bottom, hill slope, forest, and ridge were not sampled regularly, in an effort to be able to draw good general maps of climatic elements over the whole country. In the 1943 article, Thornthwaite and Leighly said that

> the standardized observations for the purposes of synoptic meteorology or for general climatologic purposes seek to avoid the local influences of vegetation and soil as completely as possible, whereas, what is required for agricultural and biological purposes are observations near the ground in the zone where the plants actually live. . . .
>
> The most important present task is in the field of microclimatol-

ogy. For the biologist it is more important to know the pattern of climatic distribution between the ground and the tree tops or the pattern over a field or a farm than it is to know the world pattern. (Reprinted with permission from *The Scientific Monthly* 57: 460, © 1943 American Association for the Advancement of Science)

After further enumeration of the myriad of microclimatic problems that face biologists, agriculturalists, entomologists, and even heating and cooling engineers, Thornthwaite and Leighly went on to propose a solution to the problem of a lack of understanding and study of climatology and especially microclimatology:

There is at present no agency in the United States that is in a position to offer the assistance asked for by scientists who, in the prosecution of their investigations, encounter climatic problems. . . . The courses in climatology that are found in many universities and colleges have usually been established and are usually maintained by departments of geography. Geographers have been interested in the science of climatology almost exclusively as a key to the regional differentiation of the earth and consequently have concentrated on descriptive climatology and on the classification of climate. . . .

The fact that climatology has in the past been largely descriptive and is not now developed to a point where it can contribute to the solution of the many urgent practical problems in agriculture and biology is in part a consequence of the departmentalization of knowledge in colleges and universities. Facility in the development of instruments for measuring the climatic elements demands knowledge of physics and mathematics; and facility in the analysis and interpretation of climatic data, once they are obtained, requires a knowledge of statistics and meteorology. But biologists do not ordinarily study physics, mathematics and statistics. Geographers, who have a very real interest in climatology, usually study neither physics and mathematics on the one hand nor biology on the other. . . .

It appears, therefore, that there is a place in the United States for a new agency, an institute for climatologic research. Such an institute can scarcely come into existence fully formed and fully armed, but must attain its final character gradually through the work it accomplishes. It would be sufficient as a beginning to have assurance of the continuous support for several years of a staff of two or three professional workers and their clerical assistants. Even ultimately the permanent staff would not need to number more than five or six.

The institute should be affiliated with a first-rate university which includes an agricultural college and experiment station. This would provide its first and most important connection. . . . Most of the research undertaken would be collaborative; and while that work was in progress representatives of agencies situated elsewhere might be temporarily in residence at the institute. Members of the institute's staff should also be moderately foot-loose, so that they might visit other institutions when occasion arose. . . .

An exceedingly important task of the institute would be the training of climatologists. This consideration is a further powerful reason why the institute should be articulated with a good university. The training offered would be on the graduate level, as is the professional training in meteorology. . . . An attempt would be made, however, to avoid overspecialization. Since the institute would emphasize collaborative research, students would be expected to acquire a rather broad acquaintance with other sciences. (Reprinted with permission from *The Scientific Monthly* 57: 463–465, © 1943 American Association for the Advancement of Science)

Although this recognition of the need for an institute of climatic research was published in 1943, it was not until 1946 and especially 1947 that, almost serendipitously, the pieces needed for its creation began to fall into place.

The Climatic and Physiographic Division of the Soil Conservation Service, of which Thornthwaite was the chief, had already been rendered largely inoperative by the departure of its staff into the armed forces or into wartime civilian agencies. However, Thornthwaite had been invited to make several trips to Mexico to advise the government on the water requirements of sugar cane and other crops in order to increase agricultural production to aid the war effort.

With a decreasing interest in the Soil Conservation Service and increasing communication with John (Jack) Seabrook, the general manager of Seabrook Farms Company in southern New

Jersey, Thornthwaite began to consider the possibility of a significant change in his life. Seabrook, a Princeton-trained officer of a large producer and packager of frozen vegetables (possibly the world's largest at the time) had written to Thornthwaite concerning the need for advice on the irrigation of orchards. Thornthwaite, in his careful, methodical way, quickly realized that Seabrook Farms had possibly the largest acreage of irrigated agricultural land east of the Mississippi River. It would be a place to experiment with some of his ideas concerning the water requirements of different crops. At the same time Seabrook Farms appeared to be a first-rate agricultural facility dedicated to solving the problems of climate and agriculture. The time was right for a move, and even though he was forty-seven years old at the time, he decided to take temporary leave from the security of federal employment to try his hand at consulting with a private agricultural producer.

Thornthwaite described some of the reasons for his move and his early experiences at Seabrook Farms in a letter dated July 8, 1946, to his old friend Homer Shantz.

> I am hard at work but there have been many changes since I last saw you. You will see by the letterhead that I am not now with the Soil Conservation Service. For several years, both in Mexico and in Washington, I have been studying drought and supplemental irrigation to combat it. Last fall I was in Mexico working with the Irrigation Commission and recommended a research program for them on water use by vegetation. The first of January they got a large appropriation for the work and last March and April I was down there again supervising the initiation of the program.
> Immediately on my return to Washington I was invited to come over here [Seabrook Farms] to study irrigation problems on this farm and to work out a system for irrigating at the right time with correct amounts of water. I took leave from the Department and am here for the summer. This is a very large establishment; 21,000 acres of which 1,300 are irrigated. Irrigation is to be rapidly extended. There

are numerous other problems here that have their solution in climatology. All in all it is the most stimulating experience of my career.

My work in the Department is going just the way yours did a generation ago and I do not intend to return to the Soil Conservation Service. I will try to make a living either as a private consultant or with my climatic institute in a university somewhere. (You have probably forgotten my article in *Scientific Monthly,* November, 1943, describing the institute.)

Shortly after Thornthwaite's arrival at Seabrook Farms as an irrigation consultant in 1946, a second piece of the institute for climatic research fell into place. During the fall and winter of 1946, Thornthwaite and his colleague at the Soil Conservation Service, Benjamin Holzman, began to develop ideas for an ambitious micrometeorology research project to be carried out at Seabrook Farms. Preliminary discussions with Jack Seabrook evoked a favorable response and a willingness to allow Thornthwaite to use Seabrook fields for the installation of sensors for profile measurements of wind, temperature, and atmospheric moisture. Although Thornthwaite still had some academic ties with the University of Maryland, the initial proposal (in March 1947) called for the work to be done at Seabrook, N.J., under contract from the Air Weather Service, with administrative support to be provided by some university still to be selected.

The original proposal called for the Air Weather Service (AWS) of the US. Army Air Force to assign a qualified officer (with a rank up to that of colonel) to Seabrook Farms to plan and organize the project and, later, two or more enlisted men to take observations and service equipment. AWS would also provide a Jeep for its personnel to use in getting around the facility as well as items of standard meteorological equipment. Thornthwaite agreed to provide technical advice to help on all phases of the project. Thornthwaite would also be responsible for obtaining the necessary space and observation sites at Seabrook Farms for experimental purposes; the AWS and Seabrook Farms would have equal access to the data obtained.

The overall purpose of the proposed research was to develop new methods of micrometeorological observations to aid in the solution of various problems of interest to both the Army and Seabrook Farms. As far as the AWS was concerned, these micrometeorological problems involved evaporation and evapotranspiration as they might influence the movement of vehicles over unpaved surfaces; weather and hydrologic forecasting; bioclimatic problems; temperature, wind and humidity observations, both in the vertical and horizontal near the ground, as related to chemical warfare forecasting; frictional effects on the general circulation; evaluating new micrometeorological sensors; and local forecasting of extreme temperatures, fog, and wind. The proposal was submitted by Thornthwaite to the AWS on March 11, 1947.

In mid-April, General Yates of the AWS wrote to Thornthwaite that the "cooperative AAF [Army Air Force]–Seabrook Farms Microclimatology Project is quite acceptable to the Air Weather Service. Initially, I wish to proceed with the project on an informal basis, with the understanding that, if it appears mutually desirable, a contract be considered at a later date." The U.S. Air Force preferred to have this informal agreement administered through a nonprofit group rather than through Seabrook Farms, a profit-making corporation. Thus, it was necessary to find a university willing to undertake the administrative tasks associated with the contract and to pay all bills. Because Thornthwaite's friend, Isaiah Bowman, was president of the Johns Hopkins University, it seemed reasonable to consider the possibility of writing the contract through Johns Hopkins. Thornthwaite met with both President Bowman and Professor George Carter, then chair of the university's Department of Geography, and it was agreed that Johns Hopkins would serve as the sponsoring agency and that Thornthwaite would take part in the geography program as a professor of Agricultural Climatology.

Further negotiations proceeded extremely slowly; on July 25, 1947, Holzman wrote informally to Thornthwaite, "It seems that I stirred up a hornet's nest around the Air Weather Service because of my screaming about the lack of suitable field instruments for measuring gradients of wind, temperature and humidity.

I pointed out that in the event a research contract was arranged with Johns Hopkins, you should be able to call upon the Army for suitable instruments to conduct the research. As a result, Yates ordered that a high-priority project be established with the Signal Corps to develop the required instruments."

On July 28, Thornthwaite notified President Bowman that Colonel Holzman had also asked for suggestions for another research proposal but that he (Thornthwaite) would not care to start on another project until the first project was well under way. Finally, a letter dated December 13, 1947, from Johns Hopkins to Thornthwaite indicated that the proposed contract would be signed by the Air Material Command, Watson Laboratories, Red Bank, N.J., with the contract to start on January 1, 1948.

As finally agreed, the research work to be carried out at Seabrook Farms by Thornthwaite (without the assistance of an Air Force officer or enlisted men) involved

the investigation of the climate and meteorology of the layer of air near the ground (within 30 feet of the surface) as influenced by soil, slope, and type of vegetation cover; the turbulent diffusion in the lower atmosphere and the determination of the coefficient of turbulent mass exchange (the Austausch coefficient), and its variation with height through the day at various seasons of the year over different types of surface; and the transport of moisture, heat, momentum, and perhaps particulates and gasses and aerosols, dust, smoke, CO_2 in the turbulent layer and through the boundary layer beneath; and the investigation of heat transport to and from the ground surface, to make a comprehensive survey of the radiation balance (incoming solar radiation and outgoing radiation) over a variety of types of bare and vegetation-covered ground; and the determination of the albedo of various kinds of land surfaces.

It is proposed that a micrometeorological field station be established to make simultaneous observations at various representative locations over a long period of time of the vertical gradients of temperature, atmospheric moisture, and wind. The location would be selected so as to include a variety of surfaces, soil types, and types of vegetation cover, and to include swamp land, water bodies and strand as well. The micrometeorological net would include a limited number of fixed stations (four to six) where detailed continuous records of temperature, moisture and wind are obtained at ten to fourteen different elevations within thirty feet of the ground, and one or more mobile stations by means of which these measurements may be obtained within the vegetation cover in different crops and types of vegetation at different stages of development. Continuous records of soil temperature and soil moisture would also be obtained. A series of evapotranspirometers would be installed and operated at one of the fixed stations to provide observational data to serve as checks on the various formulas for determining evaporation and transpiration. Daily maximum and minimum temperature would be determined at four or more levels within six feet of the ground in a large number of locations, representative of different types of exposure, different soil types and different kinds of vegetation cover. (Proposal for research between Air Material Command and C. W. Thornthwaite, Dec. 1947, in Thornthwaite files)

The proposed first-year budget for this work was $60,500, a considerable sum in 1948. Of this, $29,000 went to salaries, with Thornthwaite receiving $8,000 for full-time participation in the work. Budgeted travel came to $2,000, materials and equipment to $15,000, and overhead (at 50% of salaries) to $14,500.

In a letter to Jack Seabrook in mid-January 1948, Thornthwaite discussed the negotiations concerning overhead. He

Jack Seabrook, Warren Thornthwaite, Maurice Halstead, and Russ Mather *(left to right)* outside the Laboratory of Climatology building at Seabrook Farms in 1953. (Photography by E. Taubert, reproduced with kind permission of the Seabrook Farms Co.)

wrote, "The University held for a 50 percent overhead and the contracting officer insisted on nothing more than 25 percent. After a horrible month of stalemate they finally agreed on 28.5 percent last week." He noted to Seabrook that the change in overhead rate would not be a reduction in total funds but, rather, that he would have that much more to use in the research at Seabrook Farms.

In addition to allowing Thornthwaite to use selected agricultural fields for microclimatic observations as well as a field near the house of C. F. Seabrook (the founder of Seabrook Farms) for the installation of a battery of evapotranspirometers, Seabrook provided a building (a former railroad station near the factory yard) for research offices and Thornthwaite's extensive personal library and gave him full use of a pickup truck and a small roadster with a rumble seat to move equipment and researchers around the farm area. The Laboratory of Climatology—the institute of climatic research described in the 1943 article by Thornthwaite and Leighly—had finally been achieved. The laboratory was indeed associated with a first-rate university (Johns Hopkins), it had support for several years (on a budget that was renewed annually), it had at its beginning two professional researchers (Thornthwaite and Maurice Halstead, formerly a young associate of Thornthwaite's who worked in the Weather Bureau in Hawaii), and it undertook immediately a program to train students. Three students with master's degrees were brought in to begin work on a doctoral program in geography/climatology at the Johns Hopkins University (Russ Mather [one of the present authors], Donald Portman, and Isadore Dordick). One undergraduate student from Johns Hopkins (Winton Covey) also entered the program in the first year, and Douglas Carter entered the program in the second year. Although Johns Hopkins did not have an agricultural college or an extension program, this need was at least partially satisfied by having the research facility located at Seabrook Farms with its agronomists, soil scientists, agricultural engineers, and associated research and technical personnel. The Institute for Climatic Research was on its way!

Chapter 5

Thornthwaite and Seabrook Farms:
1946–1954

When Thornthwaite arrived at Seabrook Farms in late May of 1946, the pea harvest was well under way. As Thornthwaite stood in the factory yard surrounded by trucks filled with shelled peas, he noted that the factory was operating at full capacity, yet the trucks in the yard seemed to be fixed in place. Many trucks stood there in the bright sunshine for hours, and Thornthwaite recognized that this delay would not improve the freshness of the peas. During his first few evenings at Seabrook, he was also taken to see the night harvesting operations. Giant floodlights illuminated the rows of peas as tractors hauled harvesting equipment through the fields. The pea vines that had been cut by the harvesters were loaded by conveyer belts into dump trucks, which were driven slowly beside the harvesters. It was a fascinating operation, but it raised many questions in the curious, scientific mind of Warren Thornthwaite.

The peas were being picked day and night, yet truckloads of shelled peas were sitting in large wooden crates in the factory yard for half a day or more. The factory was running at full capacity, yet it could not keep up with the rate of delivery of peas. This was the type of problem Thornthwaite loved. Although he had come to Seabrook to advise on the irrigation of orchards, he was immediately faced with a significant problem that did not even involve irrigation.

Putting other work aside, Thornthwaite threw himself into the study of the pea problem. The story is such a classic example of the Thornthwaite approach to a research problem that it deserves to be told in detail, and it was to be repeated many times in future years. Thornthwaite essentially launched a three-pronged investigation into the pea problem. First, he questioned personnel of Seabrook Farms to find out all he could about the planting, harvesting, and processing of peas. Second, he began a detailed literature review of earlier work on the relation between climate and plant development or progress toward maturity. This work involved learning what scientists in the past had accomplished in trying to forecast harvest dates from information on planting dates and the intervening climate. Third, he began to grow peas in his own experimental area so he could observe the rate of development of the plants and sample the quality of the peas that were harvested.

A literature survey was always a significant aspect of Thornthwaite's investigations. His own reprint collection totaled more than ten thousand articles, while his extensive library included volumes of climatic data as well as books on all aspects of climatology. Thus, he easily found in his collection enough to inform him about earlier work in this particular climate-related problem, the relation between climate and crop development. He was particularly interested in the work of the eighteenth-century French scientist René Réaumur, who created a thermometer scale bearing his name that is still in use in certain cheese-making operations. Réaumur was the first observer to make an exact determination of the quantity of heat required to bring a plant to a given stage of maturity. This value was obtained by summing the mean daily air temperatures above 41°F (as registered by a shaded thermometer) between one stage of development and another. Réaumur identified this sum as the thermal constant for the plant.

Réaumur's work stimulated other investigators to make systematic observations of plant development. Observations were made of climatic elements, along with dates of budding, leafing, flowering, fruiting, and leaf fall of perennials, as well as the dates of planting, flowering, and ripening of annuals. These phenological studies contributed to an understanding of the relation between climate and the geographical distribution of vegetation.

Thornthwaite's literature search introduced him to the "heat unit" system, which was used by many canners in trying to plan the harvesting of crops. Although the heat unit system was found to possess a number of serious deficiencies, it did permit the approximate fixing of the order in which individual fields should be harvested. Thornthwaite's investigations also showed that the heat unit theory was exactly what Réaumur had outlined in 1735, though later research revealed that the so-called thermal constants were not constant. At a given stage of development of a plant, the thermal constant was smaller in high latitudes than in low latitudes. In other words, less heat was required in cold climates than in warm climates to bring about a given amount of development. This was also true for a cold year as opposed to a hot year.

Growing some forty different varieties of peas in experimental plots in 1947 not only provided Thornthwaite with visual confirmation of his scientific theories but also allowed him to perform his own taste tests. It meant that he would eat too-old and too-young peas from time to time, but it also gave him many meals of peas harvested at the peak of maturity. He loved to say, with a twinkle in his eye, "The proof of the pudding is in the eating."

From discussions with farm personnel, Thornthwaite quickly learned that the bottleneck in the pea operation arose from the limited volume of peas that the factory could process in any given period. If more peas matured or were harvested than could be processed, they had to wait in trucks in the factory

Warren Thornthwaite and daughter Elizabeth discussing the development of peas in Seabrook experimental garden in 1953. (Photography by E. Taubert, reproduced with kind permission of the Seabrook Farms Co.)

yard. Thus, the key to the problem was the number of pounds of peas (and thus the number of acres of pea fields) that could be processed per day in the factory. There was no reason to harvest more peas than could be processed in the factory in a day's time and, thus, no reason to plant more acres of peas on a given day than could be processed. Although this seemed to be relatively straightforward reasoning, implementation required considerable coordination within the upper echelons of the organization. Seabrook Farms farmed more than twenty thousand acres of land divided into ten divisions, or smaller farms, of some two thousand acres each. Each division had its own farm manager, its own labor force, and its own equipment, which were essentially in competition with the other nine divisions. Each division was told how many acres of the various crops they should plant, but other farming decisions—when to plant and how to use labor and equipment, including irrigation—were left to the discretion of the division's farm manager.

To maximize profit, the division manager sought to utilize labor and equipment most effectively. This meant planting as rapidly as possible when the time to plant arrived, so that workers could be utilized fully for a short period of time and then laid off. Maximizing the use of equipment over a short time was also desirable because there would be adequate time for repairs or converting the equipment to other uses when one operation was completed.

Thornthwaite realized that development of the pea plant toward maturity depended, in large measure, on the energy available from the sun. Lacking radiation data of sufficient detail to apply over the twenty thousand acres of farmland, Thornthwaite substituted data on air temperature. However, he realized that by using only air temperature, he would have nothing more than the old Réaumur thermal constants or the more modern "heat unit," with their recognized faults. Thornthwaite felt that the approach could be improved by using a factor that expressed the plant's potential utilization of energy in the process of development. He realized that potential evapotranspiration, which expressed the water use of a field of vegetation, or of a particular plant within that field (if it never suffered from a lack of water), could ideally be this factor. He reasoned that if the plant were utilizing energy for evapotranspiration most efficiently, it should also be using it in the same way to develop from one stage to another toward maturity. He theorized that as a plant progressed from germination to maturity, it would utilize a given amount of water (the accumulated potential evapotranspiration between those two stages) as well as a fixed amount of energy from the sun (as represented by the temperature readings). He applied this theory to peas growing in his own garden and found, for example, that Alaska peas used 16.8 centimeters of water (potential evapotranspiration) in growing from germination to maturity. He later multiplied this by one hundred in order to remove decimals. He felt that, under ideal conditions, the greater the amount of energy received from the sun, the greater will be the plant's water use and the more rapid its development. The amount of water needed (the potential evapotranspiration) is an index of the amount of growth and development of the plant.

Thornthwaite also recognized that fertilization would not alter the picture. Peas that had been planted in a poor field would mature at the same rate as ones planted in a more fertile field, but they would not be as big or healthy and the yield would be less. Thornthwaite argued that, regardless of planting date, the plant would not reach maturity until it had received the required number of energy units (which he called growth units). It should be made clear here that Thornthwaite was concerned with the development of the plant—its progress toward maturity—rather than with growth or increase in physical size. However, he felt that the correct term, *development units,* would not be as clearly understood by farmers and gardeners as growth units, and so he decided to use the latter term in his plant development studies.

Growing various varieties of peas in fields around Seabrook Farms led to the recognition that each variety required a different number of growth units to reach maturity. By spring 1947, Thornthwaite was far enough along in his thinking to suggest to Seabrook management that they try scheduling pea planting on some of the farm divisions. What he had in mind was that farm managers should be told not only how many acres of peas they should plant in that year, but also when to plant them, so that no more than the proper number of acres of peas would be harvested and arrive in the factory yard for processing on any given day. Farm managers would have to reschedule their planting operations, given that the energy received by a plant in March or April is only a fraction of the energy it receives in June. If fields were planted on successive days in March or April, they would mature on the same day in June, resulting in an excess of harvested crops in the factory yard.

Recognizing the problem, Thornthwaite suggested that if different varieties of peas with different growth units were planted on the same day, they would reach maturity on different days in June. Thus a farm manager, by the judicious use of different varieties of peas, could continue to plant rapidly in spring and still schedule the harvest to satisfy the factory's processing capabilities.

Of course, the farm managers questioned the reliability of specifying a particular harvest date. What if the seasonal weather were much warmer or cooler than normal? Wouldn't that speed up or slow down the rate of development and so invalidate the agreed-upon harvest schedule? Thornthwaite agreed that it would, but he quickly pointed out that the schedules for harvest among all fields would not be changed. A cold period would delay all fields by the same amount, whereas a warm period would speed them up equally. Harvest might start a few days earlier or later, but the flow of crops into the factory would be maintained from day to day. The factory managers also wondered what would happen if rain and muddy fields prevented harvest on a given day. This problem was largely solved by scheduling one day a week on which no crops should be harvested. Thus, that day could be used to catch up with the harvest schedule if rain had prevented some harvesting during the week.

The system was tried in 1947 for peas and worked quite

well. True, harvest dates were not predicted exactly, but the harvest order was firmly established. No more peas came into the factory yard than could be processed during that day. With scheduling, it was possible to eliminate harvesting at night, as well as the need for large numbers of human and electronic sorters on the processing lines to remove immature or over-ripe peas. The quality of the processed peas increased dramatically while labor and equipment costs declined.

After the success with the pea operation in 1947, the scheduling was expanded over the next two years to include not only all crops grown by Seabrook Farms, but also crops grown by contract farmers who sold produce to the processing plant, involving some thirty thousand additional acres. To ensure compliance with the company-mandated schedule, all contracts with private farmers included the provision that if the crops were planted according to the schedule given by the Seabrook Farms, the Farms would guarantee to buy them from the producer. If the crops were not planted according to the schedule, Seabrook Farms reserved the right to refuse the produce. With this guaranteed market, all contract farmers quickly adopted the scheduling timetable produced by Seabrook Farms.

Thornthwaite next transferred his information on planting and harvest dates of the various crops onto a growth-unit slide rule, which he called a Cropmeter. The Cropmeter changed the ordinary civil calendar, in which each day has twenty-four hours, into a climatic calendar, in which the length of each day is determined by the average amount of energy for development on that day. Thus, days in July are much "longer" climatically than days in March or April. For the Seabrook area, about ten growth units accumulated each day at the end of March, whereas the peak rate of accumulation of growth units reached twenty-three units in late July. After that, days became shorter in terms of growth units until the end of the growing season. To determine when to plant in order to har-

Warren Thornthwaite illustrating the use of an early model of the Cropmeter in the library of the Laboratory of Climatology, Seabrook, N.J. about 1952. (Photography by E. Taubert, reproduced with kind permission of the Seabrook Farms Co.)

vest at a specified time, it was necessary merely to place one of the pointers on the Cropmeter on the desired harvest date and to count back the predetermined number of growth units to maturity to obtain the proper planting date.

Thornthwaite also recognized the serious shortcoming of most seed company advertising, which specified "days to maturity" for the different varieties. He often pointed out that, with regard to plant development, days in July were not equal to days in April, and that Alaska peas planted on March 1 matured in 96 days, whereas those planted on June 15 matured in only 34 days.

During the early 1950s, Thornthwaite tried to persuade seed companies to include the idea of "growth units to maturity" rather than "days to maturity" in their seed catalogs. He believed that such information would be helpful to the individual farmer or small gardener, and he also recognized that the use of growth units would promote the sale of his Cropmeter. In cooperation with his brother-in-law, Floyd Slentz, Thornthwaite hoped that the Cropmeter could be merchandised to small farmers and local gardeners all over the country and thus provide a sizable and steady income. The key to this strategy would be the adoption of the growth unit scheme by one or more seed companies, but such support was never forthcoming.

No reason for this lack of interest was ever stated; however, it appeared that the seed companies did not want to require buyers of one or two packets of vegetable seeds to spend additional money to purchase a Cropmeter and to take the necessary time to understand its operation. "Days to maturity" would work quite well if the seeds were planted at the proper planting time, and the concept was simple to understand. An issue of the *Ford Almanac* did include a full description of the growth unit system and mentioned the availability of the Cropmeter, but sales never amounted to much. Thornthwaite tried many strategies to push adoption of the approach, but it was one case in which he overestimated the interest of farmers and gardeners.

When faced with the problem of scheduling the planting and harvesting of other garden vegetables, Thornthwaite undertook a detailed study of the literature. This led him to a greater understanding of the age-old field of phenology and a new realization of how it might lead to a revitalization of agricultural climatology.

Phenology may be defined as the branch of science that

studies periodic phenomena related to climate in the vegetable and animal world. Although phenological observations had been made for hundreds of years (e.g., records of the activities of silk worms in China), most phenological records consisted merely of dates of planting, fruiting, budding, flowering, and leaf fall of different types of plants or of particular events in the lives of animals. Studies correlating these events with various climate factors had been limited by the fact that many days elapse between successive stages. With weather factors varying from day to day during this period, the precise influence of temperature on plant development, for example, is concealed. Thornthwaite reasoned that if daily observations of plant development were available, correlations between climatic factors and changes occurring in the plant would be revealed. Because his initial interest at Seabrook Farms was with the pea harvest, it was only natural that he began to spend considerable time studying the development of the pea plant.

Thornthwaite found that the English garden pea plant was uniquely qualified for daily observation. It grows rapidly from a single growing point with little or no branching. As the stem elongates, successive nodes develop that are easily recognizable as places where the leaves occur. Each node represents a definite stage of plant development. Also, the development from one node to the next involves a series of readily discernible changes in the plant, so that it is possible, at any time, to say what fraction of development has occurred between nodes. By numbering the nodes serially, the particular stage of development on any given day can be stated precisely with a single figure. Because the pea plant grows rapidly, there is enough development in a single day to permit correlation with the weather factors of that day.

Daily observations on a number of pea plants in a field (carefully tagged so that successive readings were always made on the same plants) were started in 1947 by Thornthwaite and his

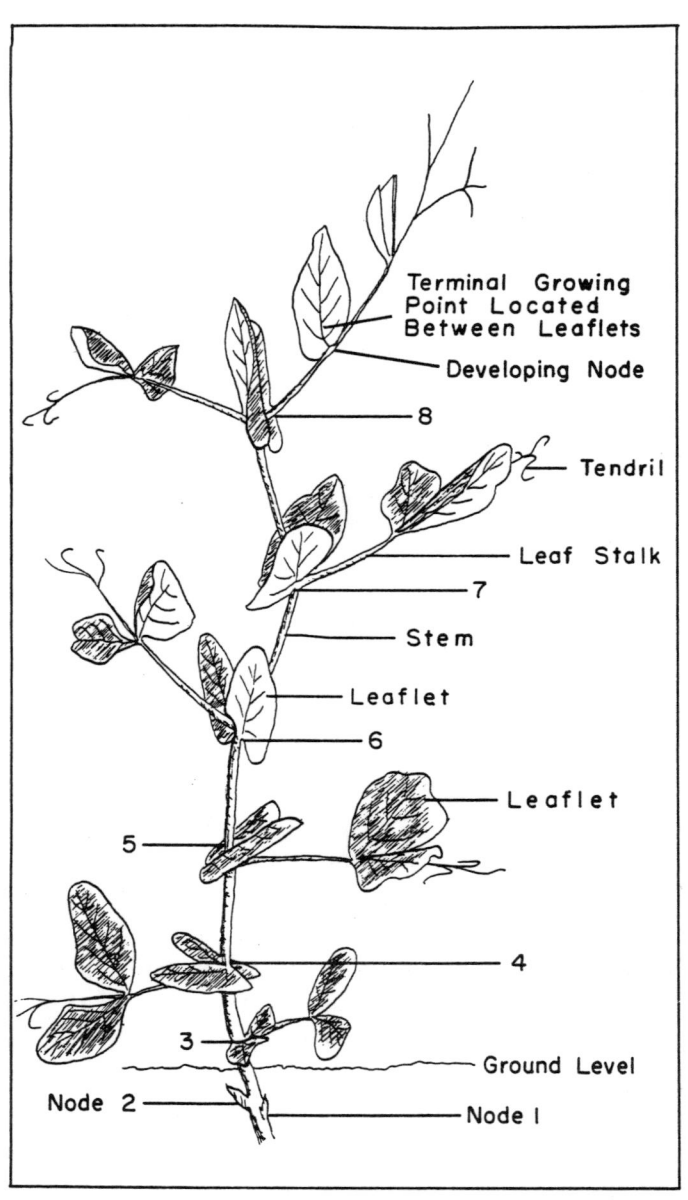

Vegetative development of the garden pea (development stage 8.8 nodes; from Higgins 1952).

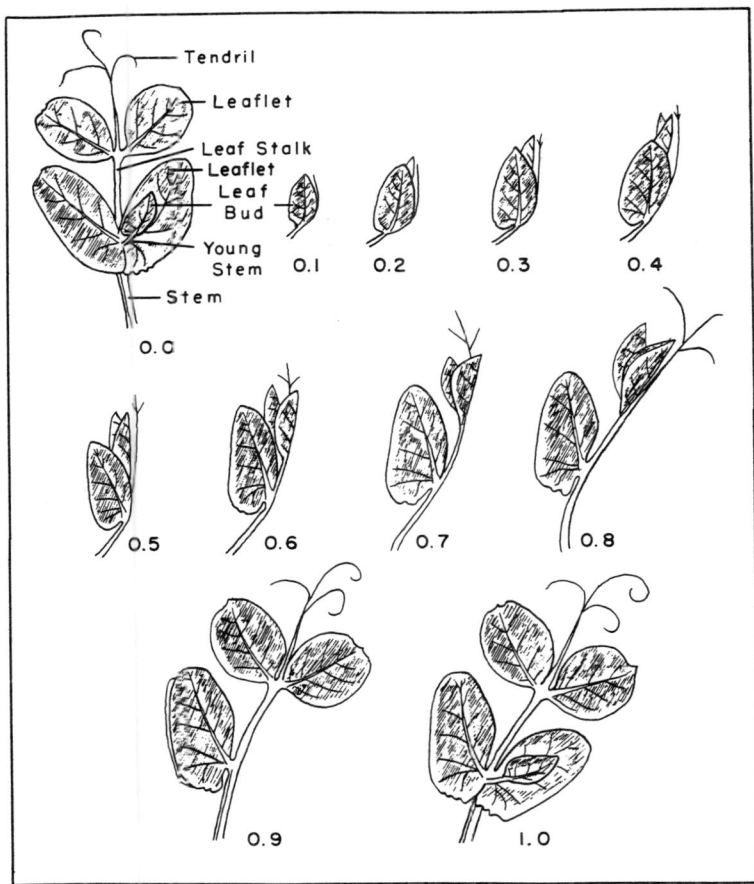

Stages of development between nodes in the garden pea, (from Higgins 1952).

daughters. Later, other workers at the Laboratory of Climatology were enlisted to make observations in order to see whether the readings were a function of the observer or if similar results could be obtained from a wide range of investigators. Successive plantings were done through the growing seasons so that plants at different stages of development at the same time could be studied.

Thornthwaite evaluating the development pattern of corn growing in Seabrook experimental garden, about 1952. (Photography by E. Taubert, reproduced with kind permission of the Seabrook Farms Co.)

Thornthwaite was beginning to appreciate what daily plant readings could do for agricultural climatology. All work in the field to this time had related plant development to different weather factors such as solar radiation, wind, humidity, and temperature. However, the plants were not growing in the same environment where the weather instruments were placed and therefore were reacting to somewhat different conditions. Also, it was not possible to determine the integrated effect of different weather elements on the plant, for this would vary from one species to another and might even vary through the growing season for a single species. Thornthwaite suggested that the plant itself be used as the integrative instrument—that is, to make daily observations of the plant and, thus, to read it as an instrument that integrated temperature, wind, humidity, solar radiation, and other weather factors. This was a revolutionary concept in the early 1950s, one for which the field of agricultural climatology was not prepared.

The failure of researchers to adopt the idea of using the plant as an integrating tool was, in large measure, attributable to a lack of any real understanding of how such a measurement could be used. Even Thornthwaite's own understanding of the significance of such daily plant development information came slowly and only after he had examined the results from an expanded pea observation data network. In 1952, Thornthwaite began to enlist the help of volunteer observers worldwide. The volunteers were furnished with seed, markers, instructions, and observation forms. In the first season, observations were made in Illinois, New Jersey, Louisiana, Wisconsin, Massachusetts, Florida, and Ontario, Canada. By the second year observers in Europe, Asia, and Latin America had been included in the program. Preliminary analyses of the Seabrook observations appeared in 1954. A paper prepared by Thornthwaite and Mather in that year shows the generally similar relation between pea node development and potential evapotranspiration.

Thornthwaite felt that the close relationship between potential evapotranspiration and daily plant development could be used to gage the health of the plant. In other words, if the computed value of potential evapotranspiration, which combined both energy and moisture factors, showed that four-tenths of a node should have developed on a given day but observations

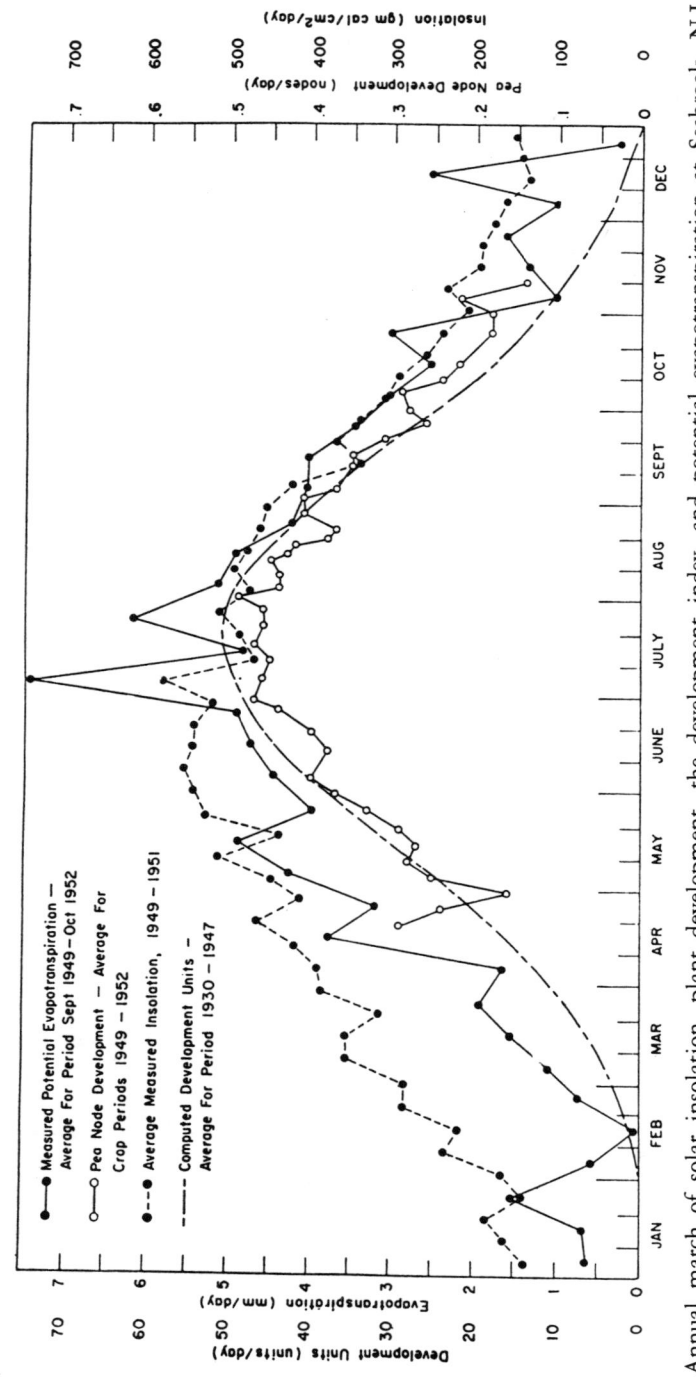

Annual march of solar insolation, plant development, the development index, and potential evapotranspiration at Seabrook, N.J. (reproduced with permission of the Meteorological Society, from Thornthwaite and Mather, *Meteorological Monographs* 2, no. 8 [1954]:8.

on the plant itself showed only two- or three-tenths development, this would be an immediate indication that the plant was suffering in some way. It clearly had not been able to utilize the energy from the sun at the optimum rate to develop to the extent that the climate would allow. The farmer could take this as a warning sign, before any wilting or reduction of yield would be evident, and do what was necessary (e.g., fertilize or irrigate) to bring the rate of plant development up to the potential rate suggested by the computation of potential evapotranspiration.

This concept of using the plant itself as an integrating instrument and of reading the plant daily as one would a thermometer or psychrometer was many years ahead of its time. Thornthwaite and his colleagues at the Laboratory of Climatology devoted about eight years of research time to this study of phenology, but eventually they were forced to give it up because of financial reasons, not because of any perception that the work was of little value to agricultural climatology. The Laboratory of Climatology was supported entirely by research grants from the U.S. Air Force, Office of Naval Research, Army Signal Corps, Arctic Institute, Resources for the Future, and Army Corps of Engineers, among others, and these groups had little interest in supporting phenological studies. The initial work was supported by Seabrook Farms, but when Thornthwaite severed his connection with the Farms in 1954, no sponsor for the phenological work could be found. This was a great disappointment to Thornthwaite, for he believed that the new phenology would revitalize the old field of agricultural climatology and aid all phases of the farming program.

While he was developing the growth unit approach to crop scheduling, Thornthwaite also recognized that climate affects transpiration (the transfer of water from vegetation to the air) in the same way that it affects plant growth and development. This suggested that irrigation could be scheduled in the same manner as crop planting and harvesting. Because Thornthwaite was originally brought to Seabrook Farms to advise on the irrigation of orchards, he lost little time in trying to determine the water needs of various crops. To do so, he installed a battery of six evapotranspirometer tanks in the fall of 1947. These were large (eight feet in diameter and about two feet deep), soil-filled tanks in which a particular crop was planted and which were placed in a field having the same vegetation cover. The soil in the tanks was kept moist by maintaining a water table at about sixteen inches depth. Similar tanks had been designed for use in Mexico during World War II to determine the water requirements of various agricultural crops, but because of conditions beyond Thornthwaite's control, they had not provided any information on plant water need until after the Seabrook tanks had been put into operation.

The six Seabrook evapotranspirometers were operated continuously until the mid-1950s, when Thornthwaite terminated his consulting arrangement with Seabrook Farms and the Laboratory of Climatology had become the Drexel Institute Laboratory of Climatology. Three more evapotranspirometer tanks were later installed at Centerton, N.J., and were operated for about a decade until leaks in the tanks made it impossible to obtain reliable information on plant water use.

From his own evaluation of the results of so-called pot experiments, in which researchers had grown grasses and crops in isolated pots or garbage cans exposed above ground, Thornthwaite had already concluded that it was unsound to compare the water use of different crops growing in adjacent tanks. Tall-growing crops are exposed to more solar energy than are short-growing plants, and this oasis-like exposure results in overestimation of the true climatic demand for water. Similarly, irrigation of the soil in the tanks, but not in the surrounding area, keeps the moisture content in the field tanks above that of the surroundings and again, an oasis-like exposure develops. A wet area in the middle of a drier field results in the transfer of energy to the wet oasis by advection, as well as by solar radiation, and again results in an overestimation of evapotranspiration.

Thornthwaite continued to check, test, and modify his definition of potential evapotranspiration. Although it was originally defined in 1944 as "the water loss that would occur if soil moisture were constantly at the optimum level for maximum plant growth,' he soon recognized the need for more detailed requirements. He outlined fairly strict specifications for the operation of the soil-filled tanks used to obtain measurements of potential evapotranspiration and he wrote, "A well-mixed sandy loam having a moisture equivalent about 18% and wilting percentage about 8% should be used in all locations. . . . The depth of ground water most satisfactory for plants is approximately 50 cm. At greater depths the plants are deprived of needed water, and at lesser depths evaporation becomes excessive and plant roots are restricted" ("Report of the Committee on Evaporation and Transpiration, 1945–46" p. 722).

Most of the guidelines were followed in setting up the evapotranspirometer installation at Seabrook Farms. Although the need for moist conditions both within the field tanks and in the surrounding area was not mentioned in the 1946 article, a short period of observations early in 1948 at Seabrook made it clear that failure to keep moisture conditions similar over the whole experimental area resulted in excessively high values of measured evapotranspiration from the tanks. Thornthwaite recognized the need for vegetation in the tanks to be similar to that in the surrounding area, and although he had recommended the use of native grasses or salt grass in his 1946 report, he decided to use a variety of agricultural crops in the Seabrook installation so that the results would be more immediately useful in his irrigation consulting work. Thus, over several years and a number of cropping seasons, the tanks and surrounding experimental areas were planted with peas, beans, sweet corn, spinach, and finally turf grass.

The most significant result obtained from the evapotranspirometer installation was the recognition that under potential evapotranspiration conditions—which meant that the soil was maintained at field capacity so that the plants had all the water they needed—the type of vegetation had little influence on the rate of removal of water from the soil. On any given day, peas were found to use as much water as spinach, spinach as much as beans, beans as much as corn, and corn as much as turf grass. Although these were the only crops grown in the evapotranspirometer tanks, Thornthwaite reasoned from these observations that the water use of all crops would be the same under potential evapotranspiration conditions. This was significant because it eliminated type of vegetation as one of the possible controlling factors in potential evapotranspiration, and it verified his early suggestions to Homer Shantz that under moist conditions, the type of vegetation cover was unimportant in controlling plant water loss.

In 1954 Thornthwaite achieved an interesting verification of this result during a multiagency expedition to O'Neill, Nebr., in which Laboratory of Climatology scientists participated. Using data on soil moisture content, solar energy, and moisture removed from the soil under short grass prairie, it was found that approximately ninety percent of the energy from the sun went into evapotranspiration when the soil moisture was at field capacity. When the soil moisture content was below field capacity, a proportionately smaller amount of solar energy was utilized in evapotranspiration. One could thus reason that when the soil was moist, nearly all available solar energy was utilized for evapotranspiration, and that potential evapotranspiration would not vary significantly from one crop to another.

A second verification came from Toronto, Canada, where one of the authors, Marie Sanderson, grew timothy and crested wheat grass in evapotranspirometer tanks. Timothy was known to be a plant that needed moist conditions and would not do well under conditions of limited moisture supplies. Crested wheat grass, a hardy grass imported from Russia before the turn of the century, was known to survive under very harsh conditions. When timothy and crested wheat grass were grown in the same vicinity in evapotranspirometer tanks, Sanderson found that, under conditions of always-adequate soil moisture, both crops used the same amount of water. Probably the crested wheat grass would have survived with much less water than would the timothy, but when both crops could use all the water they wanted, they used the same amount, based on the available energy from the sun.

It took some time for farmers and agricultural researchers to understand and accept this conclusion. Their records always showed that the deeper-rooting corn, with a longer growing season, used more water than did peas or spinach grown in the cooler spring or fall seasons and for shorter time periods. And certainly their findings were correct. What Thornthwaite was trying to teach them was not that crops under variable soil moisture conditions would use different amounts of water, but that with soil moisture conditions at field capacity, the water use on a particular day depended on the available energy from the sun rather than on the type of vegetation. Growing peas and corn on the same day under conditions of adequate soil moisture resulted in both crops using the same amount of water. That farmers would not normally grow both crops at the same time was immaterial; the conclusion was still valid.

This finding resulted in a further refinement of Thornthwaite's definition of potential evapotranspiration to include the requirement that the vegetation cover be homogeneous so that a uniform crown area with a closed cover would be presented to the incoming solar radiation. This latter requirement ensured that bare soil areas would not appear between the rows of vegetation. Thus, potential evapotranspiration was defined as the water loss from a closed, homogeneous cover of vegetation that never suffers from a lack of water. Recognizing the problem of reflection of radiation from vegetation, Thornthwaite later specified that the albedo of the vegetation cover should be between twenty and twenty-five percent and preferably between twenty-two and twenty-three percent. This ensured a fairly constant amount of absorption of radiation by the plant and its ultimate use in evapotranspiration. Thornthwaite's empirical expression for potential evapotranspiration was verified by evapotranspiration experiments in other parts of the world—Mexico, Israel, Nigeria, Ireland, Hong Kong, and Argentina—and the results were included in publications of the Laboratory of Climatology.

Thornthwaite and Mather (1954) wrote in "Climate in Relation to Crops,"

The distinction between potential and actual evapotranspiration can be made clear if one considers the sparse vegetation of the deserts which requires little water to sustain life. That the xerophytic plants of the desert are able to survive with very little water does not mean, however, that more would not be used if it were available. Thus, there is a distinction between the amount of water that actually transpires and evaporates and that which would transpire and evaporate if more water were available. Where the water supply is increased, as in an irrigation project, evapotranspiration rises to a rate that depends only on the climate ("Climate in Relation to Crops," 1954, p. 4, with permission of the American Meteorological Society).

Thornthwaite's work on potential evapotranspiration permitted the development of a simple but effective irrigation scheduling system. Many farmers measured rainfall at a number of places on their farms, but few had any information on the plant water losses (the return flow of water to the atmosphere). Thornthwaite explained that irrigation scheduling could be followed using a simple bookkeeping procedure to keep track of moisture losses through evapotranspiration. Losses not promptly replaced by precipitation would have to be satisfied by irrigation.

Such a system prevents overirrigation—and thus leaching of nutrients through the soil—and ensures that the soil moisture content is not deficient so that plants never show any signs of distress. Thornthwaite described the water budget bookkeeping procedure as similar to bankbook accounting: periodically, water (money) is added and, at other times, water (money) is removed from the account. The bank balance is analogous to the soil moisture condition.

When Thornthwaite proposed this irrigation bookkeeping system to Seabrook managers in the early 1950s, they were skeptical. Although willing to let Thornthwaite experiment with his bookkeeping approach, they relied on a semiautomatic system for determining soil moisture content by weighing and then drying samples of soil. For one spinach-growing season in the early 1950s, the Seabrook soils laboratory sent personnel into a spinach field daily to obtain samples of soil from different locations. Carefully sealed, these samples were taken to the laboratory, where they were weighed, dried in an oven, and then reweighed to determine the amount of water in each sample. These data were then averaged to obtain information on the amount of water removal each day by evapotranspiration as well as the actual balance of moisture in the soil of the spinach field. The operation involved one full-time technician each day.

Meanwhile, Thornthwaite sat at his desk in the farm office and used the observations of the mean temperature of the previous day and the total precipitation, if any, to calculate the average soil moisture balance in the same field. It would take him no more than two minutes to convert the temperature data to potential evapotranspiration and then to determine precipitation minus potential evapotranspiration for the day. He would then subtract this number from or add it to the soil moisture figure obtained the day before to obtain the new value of average soil moisture content in the field. When this value dropped below some predetermined value, indicating that the plants were beginning to suffer from a lack of water, he would call for irrigation on the field. The quantity of irrigation would be established by the field's water-holding capacity and its actual soil moisture content. Irrigation should not remoisten the field above field capacity because, if this were to occur, nutrients and fertilizers would be leached out of the upper layers of the soil.

Thornthwaite was pleased, but not at all surprised, when it turned out that his daily soil moisture computations and the daily measurements differed hardly at all. He did not fail to point out that his work required two to three minutes, whereas the soil sampling, weighing, and drying required most of a day. He loved to describe himself as a "Fanny Farmer"—meaning that he could sit in his office, schedule both the planting and harvesting of various vegetables, and calculate their irrigation needs without ever having to go outside. Of course, he did go out, for he was an avid field researcher, but he liked to call attention to the value of applying the scientific method to the problems of agriculture.

Thornthwaite had scarcely started his consulting work for Seabrook Farms before he was asked for advice on other aspects of agricultural climatology by Jack Seabrook, the youngest son of C. F. Seabrook. C. F. Seabrook's three sons managed different aspects of the company's operation under the overall direction of C. F. himself. While Jack handled the farming operations, another brother oversaw the processing plant and freezing operations, and the third brother was in charge of sales and distribution of the product.

In attempting to increase yields, Jack Seabrook asked Thornthwaite how more detailed weather forecasts might be utilized in the farming operation. Thornthwaite's somewhat cynical answer expressed his own lack of enthusiasm for the activities of the U.S. Weather Bureau:

> Three types of weather information are needed for improved farm operation: (1) short-range weather forecasts (a few hours to a few days), (2) long-range weather forecasts (a few weeks to a season), and (3) climatic and microclimatic analyses.
>
> It is logical to look to the Weather Bureau for aid in making better adjustments to the weather. Actually, no real help can be obtained from the Weather Bureau. In the summer of 1943, Seabrook Farms began receiving via Western Union from Washington two types of forecasts. Thirty-six hour forecasts came daily and five-day forecasts twice a week. The service was soon found to be of no particular value and was discontinued. The Weather Bureau now provides a three-day farm forecast. No new knowledge goes into the forecasts and they are no better today than they were three years ago. In fact, there has been no real improvement in 30 years.
>
> The Weather Bureau does not attempt a long-range forecast. Meteorologists have made exhaustive studies along every conceivable line but so far have discovered no means of making successful long-range forecasts.
>
> The Weather Bureau has no competence and little interest in climatic analysis. . . . A good example of this fact is afforded by a current Weather Bureau investigation of peas. On August 7, 1946, the Weather Bureau stated that it intends to publish "the daily heat units above a base of 40 degrees . . . as a means of estimating the rate of growth and approach toward maturity of different varieties of peas." This method was demonstrated to be unsatisfactory in 1849 and was exhaustively discussed in a Weather Bureau Bulletin of 1905, and discarded. The present investigators will presumably rediscover the inadequacy of this method the hard way. (Unpublished memo from Thornthwaite to J. M. Seabrook, September 5, 1946)

Because Thornthwaite had come to Seabrook Farms on leave from the Soil Conservation Service, where he had been chief of the Climatic and Physiographic Division, it was natural that

Jack Seabrook would ask Thornthwaite about soil conservation work at Seabrook. Jack recognized that erosion was an increasing problem and that yields had not increased as they should have under so-called improved farming techniques. This question touched a sensitive nerve, and Thornthwaite's answer to Jack was one of the most passionate that he ever made:

> The national soil conservation program is a crusade; an old time religion. All the elements of evangelical religion are present: sin, damnation and salvation. We sin by plowing and cultivating the land, but we can be saved by terracing and planting on the contour. Acres terraced are like souls saved. The high priests of soil conservation are evangelists, indistinguishable from Dwight Moody or Billy Sunday. They even have their queer sects that have special views on "sin" and special means to achieve "salvation"; raindrop or splash erosion, for example. Farm planners know almost nothing about the nature of erosion, what causes it, how it varies from season to season. They have their bag of magic, their means to salvation; terraces, etc., etc., and their faith. To question the efficacy of these conservation practices is a mark of an unbeliever.
>
> I entered the Soil Conservation Service when it was created eleven years ago, to head a Division in Washington. We have studied the history of soil conservation practices the world over and find that engineering practices everywhere have done far more harm than good. (Unpublished memo from Thornthwaite to J. M. Seabrook, October 11, 1946)

The frustration of a decade of studying the practices of the soil conservationists and the government recommendations for "sound" approaches to conservation shows clearly in Thornthwaite's letter. He went on to express a few of his own thoughts on the soil conservation measures that should be employed at Seabrook Farms:

> Soil conservation practices are only special farming practices, and too often they do not reach the seat of the trouble; like the aspirin that makes the headache disappear without touching the cause. The soil has been misused at Seabrook Farms. As a result of continued

cultivation, the soil structure has disappeared and its capacity to absorb and hold water has been greatly impaired. Likewise its air capacity (aeration), also necessary to healthy crops, has diminished. Water does not enter the soil readily and consequently during hard rains, some runs off and the soil erodes. . . . The soil [at Seabrook] is Sassafras loam. It should be easily managed, should be very permeable and should hold as much as 2.00 inches of water per foot depth. Actually it is surprisingly impermeable and suffers very much from erosion, and holds only about 1.30 inches of water per foot depth, no more than would be expected of a loamy sand. The terraces that have been constructed there do no good. They actually make the soil less permeable and reduce its water holding capacity. . . . More water than ever is lost, cultivation is made difficult and erosion is not controlled. The task is to restore soil structure and increase the permeability and water holding capacity of the soil to eliminate runoff rather than to attempt to conduct the runoff water away from the field with a minimum of erosion.

Water conservation and soil conservation are inseparable and the more thought that is given to the former, the fewer will be the problems of the latter. In the climate of South Jersey, the rainfall of the growing season does not equal growing season water need. Accordingly, crops usually suffer from drought. Some things can be done to bring water need and rainfall more nearly together. One possibility is to increase the storage capacity of the soil for moisture. Every added inch of usable water in the soil is as good as an extra inch of rainfall. A second possibility is to increase the rate of water entry into the soil to prevent loss through runoff. An inch of water saved is equal to an inch of rain. A third possibility is to prevent or reduce loss of water from the soil surface by evaporation. A fourth is to reduce the water needs of the crop. A fifth possibility is to make up the water deficiency by use of supplemental irrigation. To the extent that these water conservation measures reduce runoff, they will prevent erosion and thus are soil conservation measures as well. (Unpublished memo from Thornthwaite to J. M. Seabrook, October 11, 1946)

The above memo illustrates the practical nature of Thornthwaite's research at Seabrook. That Thornthwaite's work was highly regarded by Jack Seabrook is evident from an article the

Warren Thornthwaite adjusting the exposure of thermocouples to obtain air temperature profile measurements over an experimental site near Seabrook, N.J. in 1953. (Photography by E. Taubert, reproduced with kind permission of the Seabrook Farms Co.)

Thornthwaite firing a smoke puffer to determine air turbulence at Seabrook airport. (Photography by E. Taubert, reproduced with kind permission of the Seabrook Farms Co.)

latter wrote in 1953 for *Weatherwise:* "The Seabrook laboratory has pioneered the field of agricultural microclimatology, and with the precise observations that we obtain in the layer near the ground, we expect gradually to come to understand the climatic factors that affect our farming operations" (Seabrook, 1953, p. 37).

Thornthwaite was also interested in basic, theoretical research. One of the basic research tasks he envisaged for the Laboratory of Climatology was to obtain a workable, physically sound, mathematical expression for the variation with height of temperature, moisture, and wind in order to learn more about the phenomenon of turbulence near the ground. While considering this problem, Maurice Halstead, assistant director of the laboratory, happened to see a Camel cigarette billboard in Times Square, N.Y. The smoker portrayed on the billboard was blowing smoke rings that would drift over the viewers' heads in the busy New York intersection. Viewing the smoke rings at different times, Halstead realized that the length of time the rings remained visible varied during the day in the same fashion that atmospheric turbulence varied.

Upon returning to Seabrook Farms, he reported these observations to Thornthwaite, who not only agreed that smoke rings could provide a measure of the changing turbulence during the day but also saw an application of the technique that could solve a perplexing problem in agricultural climatology. Seabrook Farms had several airports from which crop duster planes flew to apply insecticides to the field crops. The pilots knew that the most effective applications of insecticides were carried out in the early morning or late afternoon hours. The reason, of course, was the general lack of turbulent diffusion at those times, so that the dust from the airplanes would slowly settle down on the plants. If the pilots dusted later in the morning or in early afternoon, they knew that they would have to increase the rate of application in order to achieve a satisfactory coverage, for turbulence would disperse the dust as it settled out of the air. Thornthwaite quickly saw the value of developing a small "smoke puffer" device, which would shoot a puff of smoke into the air so that its duration of visibility could be timed. He felt that the dissipation time of the puff would be a direct measurement of the turbulent structure of the lower layers of the atmosphere. Observers with stop watches would follow the puff visually as it moved over the ground away from the puffer. The more turbulent the air, the more rapidly the puff thinned and dissipated. Under very stable conditions, the smoke puff could be followed for up to two minutes.

The next step was to determine the relation between smoke puff dissipation time and the cover of insecticide dust on the field. This was done by placing a number of glass slides, lightly covered with Vaseline, at leaf height in the field to be dusted. Smoke puff dissipation time was then correlated with the amount of dust collected on the slides. The final step was to ask the farm manager if there had been an adequate application of dust. If the manager said that there was adequate coverage, information on the rate of application from the pilot and the dissipation time were plotted graphically to create a curvilinear relation between dissipation time and satisfactory dust application. With very short dissipation times (less than twenty seconds), very high rates of dust applications were needed to ensure adequate coverage. However, if dissipation time was over one minute, dust application rates could be lowered significantly and still give a satisfactory coverage of the field.

As the system was conceived, the crop duster pilot would be supplied with a smoke puffer and a can of black gunpowder that provided the smoke. Upon arriving at the airport, the pilot would shoot several puffs of smoke and time their rate of dissipation. This information would determine the proper rate of application of insecticide. After crop dusting, the pilot would return to the airport and repeat the process while the plane was being reloaded. Thus, the pilot would continue to adjust the application rate as the turbulent condition of the atmosphere changed.

The smoke puffer was originally conceived to help clarify some of the complexities inherent in an understanding of atmospheric turbulence. It ultimately achieved a practical utilization as a simple instrument to solve a particular problem in agricultural climatology. Unfortunately, the instrument never gained wide utilization because of state regulations restricting the shipping and storing of black gunpowder. Other materials such as powdered chalk and ashes were tried in place of gunpowder, but no readily available substances could be found that would create the same visible puff produced by a spoonful of black powder.

Another interesting and successful study that Thornthwaite carried out at Seabrook Farms involved the disposal of effluent from the processing plant. Seabrook Farms had already installed a primary treatment facility as well as a small, high-rate biofilter plant to carry out secondary treatment of the domestic effluent generated by householders in the village of Seabrook. All of this treated effluent was allowed to flow into a nearby creek that, many miles downstream, entered a recreational lake near the city of Bridgeton, N.J.

As Seabrook Farms expanded its processing operations during and following World War II, the volume of effluent from the processing plant increased. Plant effluent was sent to the primary treatment plant, but there was too much to be treated by the small secondary treatment facility. The temporary solution was to dispose of the lightly treated plant waste into the same creek that received the treated domestic waste. However, the processing plant produced many times the volume of effluent being produced from the village of Seabrook. In fact, plant effluent far exceeded the normal summertime flow of water in the small creek. When algae blooms began to form in late summer each year on the recreational lake, homeowners in the vicinity joined in a lawsuit to stop the disposal of plant effluent into the creek.

Few options were available to Seabrook Farms. It could increase the size of its high-rate biofilter plant in order to handle the ten to fourteen million gallons of water daily from the processing plant, but there was no guarantee that the biofilter would purify the effluent. The organisms within the bed of crushed stones that oxidized the organic material in the effluent might not thrive in the water from the processing plant. Also, there was not a sufficient volume of domestic waste to mix with the plant effluent to ensure a proper mixture for the organisms in the biofilter.

Many other food processing plants stored plant effluent in large lagoons to allow natural oxidation to occur. Some even encouraged the oxidation process by bubbling air through the water in the lagoons. For Seabrook, there were drawbacks to the use of lagoons: the large capacity needed to handle the great volume of Seabrook effluent, the need to remove and dispose of large volumes of dewatered sludge, and the fact that such an extensive lagoon surface might provide a breeding ground for flies and mosquitoes. Also, because of the large need for irrigation water, it was natural to think of using this wastewater for irrigation. Because Thornthwaite had been hired to advise on irrigation, management turned to him for suggestions. Based on his knowledge of potential crop needs for water, and of the limited ability of cultivated fields to absorb rain or irrigation (about one inch every four or five days in summer), he told the farm managers that they would have to irrigate more than two thousand acres of land in the vicinity of the processing plant to handle the effluent.

Management wanted to test the concept and, perhaps, to verify Thornthwaite's conclusions. Using a large, high-pressure sprinkler that discharged some twenty-seven thousand gallons of water per hour over a circular acre, it was possible to cover the sprinkled area with one inch of water every hour. Operating in a cultivated field for an eight-hour period, the sprinkler deposited eight inches of artificial rain on the field. When the operation was terminated after the first day, the area was a thick soup of mud, standing water, and battered vegetation. All agreed that such a process would not be possible on a regular basis throughout the growing season.

When Thornthwaite walked into a wooded tract adjacent to the cultivated field—which, he had noticed, had also received the sprayed water each time the sprinkler rotated—he found few signs of damage from the eight hours of spraying. Clearly, the lack of years of cultivation and harvest operations in the wooded tract had resulted in a more open, permeable soil structure. The irrigated water moved rapidly down through the soil because there was no "plow sole" or compacted layer just below the depth of plowing to impede percolation. Thornthwaite immediately suggested moving the sprinkler system a few hundred feet into the wooded tract and trying the experiment again. Over the next two days more than forty-eight inches of water were sprayed into the wooded tract, and when the area was surveyed at the end of that time there was little evidence of the tremendous amount of water that had been applied. The water had infiltrated and percolated downward through the soil.

Thornthwaite then made one of those enlightened decisions of which he was so capable, a decision involving a large financial outlay on the part of Seabrook Farms and based not on sure knowledge that it would be successful but rather on his experience and intuition. He suggested that the Farms personnel set up a complete irrigation system in the wooded tract involving fifty-four sprinklers (nine on each of six lines). To keep costs of operation as low as possible, an open two-mile long canal was to be dug from the primary treatment plant to the wooded area so that the effluent would flow by gravity to the disposal field. Six large pumping stations were located in the canal in the wooded area, each to pump over eighty thousand gallons per hour under high pressure to three sprinklers on the irrigation lateral connected to that pump. With three sprinklers on each line operating for eight hours each day and a total of nine sprinklers on each line, it was possible to operate the line continuously and still provide sixteen hours each day for each sprayed area to drain and dry. By operating all six lines for twenty-four hours a day, they could spread more than 11.5 million gallons of water over the waste disposal tract each day and no area would receive more than eight inches of water. If

Pumping station 3 on the factory effluent canal in the wastewater disposal tract, Seabrook, N.J., in 1951. (Photography by E. Taubert, reproduced with kind permission of the Seabrook Farms Co.)

more water came from the processing plant, there was enough storage capacity in the two-mile long canal to even out the peaks and valleys in the pumping schedule.

This aspect of the waste disposal operation was fairly straightforward, and the two-day test in the wooded area suggested that it would work. The question that remained was whether the spraying operation and the subsequent percolation through the soil would lead to purification of the processing plant effluent. Thornthwaite felt that this would happen, but he certainly had no guarantee that the area would be able to purify such a large amount of effluent. The need to cease disposal into

Warren Thornthwaite and Jack Seabrook discuss the operation of a sprinkler nozzle used in wastewater disposal area at Seabrook Farms in 1951. (Photography by E. Taubert, reproduced with kind permission of the Seabrook Farms Co.)

the creek was urgent, however, and so plans were quickly finalized. The canal was constructed, along with a flume to measure the volume of water flowing in the canal and one aqueduct to carry the water over a low ravine. By May 1950 the system was ready to go.

Effluent from the processing plant generally averaged 6 to 10 million gallons per day, though it sometimes reached 12 to 14 million gallons for short periods. Biological oxygen demand ran from 100 to 1000 parts per million (ppm). Such concentrations in a small creek could prove to be a problem. Total solids ran from 200 to 1200 ppm, of which dissolved solids constituted 90 percent. The effluent had 0 to 300 ppm of sodium, 2 to 8 ppm of nitrate, 250 to 450 ppm of chloride, and a pH of 4.5 to 6.5.

The total disposal site at Seabrook consisted of about 180 acres, of which fifty-four were to be used for actual spray disposal. Although the soils in the wooded tract were generally sand and sandy loam, there were clay lenses at a depth of one to two feet in some locations. Thus, it was understood that not all areas would be equally suitable for receiving the effluent and that some adjustment of sprinkler heads would be necessary to prevent gullying and erosion in areas of decreased infiltration and steeper slopes. Thornthwaite recognized that the presence of vegetation was a primary prerequisite for the successful operation of a woods irrigation system, because the vegetation cover served to maintain the infiltration capacity of the soil and to prevent erosion. In addition, it served as a water filter, keeping some suspended and dissolved solids in the sprayed effluent out of the soil pores, and, of course, it increased evapotranspiration water loss. The wooded tract selected for the work had white and black oaks as the primary tree species with flowering dogwood and mountain laurel as the principal understory vegetation. Blueberry and huckleberry were the main ground cover. The woods were fairly open and easy to walk through in most places.

Twenty-nine ground water wells were installed in the area to allow monitoring of the changes in the depth to the water table. At the beginning, the average depth to water was twenty feet, ranging from less than five feet in three wells installed in ravines and low-lying areas to more than thirty feet in five wells in higher areas.

From May to December of 1950, some 1.14 billion gallons of effluent were sprayed onto the disposal tract, in addition to some 185 million gallons of water received in the form of precipitation. Because of variations in operational procedures and the shifting of line and nozzle positions during the year, the amount of artificial rainfall (sprayed effluent) applied to the forest area varied considerably from one section to another. In those regions not reached by the spray no water was applied. The maximum amount of water, 1,183 inches, which was applied to a region near the very center of the disposal area, resulted from the overlap of the spray from several nozzles. The average application at a sprinkler location was four hundred to six hundred inches during the 1950 operating season, with higher values occurring in regions of spray overlap.

Maps were prepared to illustrate the change in the depth to the water table during the first season of operation as well as the changes from the beginning of the first year to the beginning of the second year. There was no operation of the irrigation disposal system from December 11, 1950, to April 30, 1951. During spray operations, the maximum rise in the water table was found to be twenty-two feet, though the average rise was only ten feet. Little groundwater mounds were found under each of the sprinkler locations, indicating that the water from each nozzle moved fairly vertically downward with little lateral spread. Recovery of the water table to its previous level during the winter and spring periods of no disposal operations was nearly complete. The average change in depth to the water table over the entire area was a rise of just two feet from May 1950 to April 30, 1951.

Maintenance of the vegetation cover over the disposal tract was a prime concern during the first year of operation. The force of the water from the giant sprinklers did considerable damage to the bark, leaves, and branches of nearby trees, and many areas of dead trees appeared during the second year of operation. Although this concerned many of the visitors to the site (and once the word got out about the spray disposal area, visitors came by the scores), Thornthwaite was careful to point out that total vegetative cover was increasing rapidly. The sprayed effluent was rich in plant nutrients and fertilizers, and

Total depth of water applied to the wastewater disposal area of Seabrook Farms in 1950. (Photography by E. Taubert, reproduced with kind permission of the Seabrook Farms Co.)

the weeds—pokeweed, pigweed, fireweed, horseweed, and lambs quarters—flourished, though blueberries and huckleberry were greatly thinned. By the second year, some of the weeds had reached heights of ten to twelve feet and had stalks over an inch in diameter. They provided a better ground cover than was present on the area during the first year's operation.

Soil erosion became a problem during the first year because of the initial thinning of the vegetation cover and overland runoff from the less permeable areas. However, with relocation of the sprinklers and the development of more adequate ground cover in the second year, soil erosion was eliminated as a problem. The solids in the effluent remained on the ground, as did the tree leaves, and were broken down into fine humus by the organisms in the upper level of the soil. Tests of the percolating effluent revealed that most of the pollution was removed in the top two to three inches of the forest duff. In fact, the biological oxygen demand of the sprayed effluent was reduced by ninety-nine percent after just a few inches of movement through the upper soil of the wooded area. Weekly measurements of depth to the groundwater table provided detailed information on groundwater movement. Because of state and county interest in the fate of such large volumes of factory effluent, there was careful monitoring of all inflow and outflow of water in the woodland tract.

Although a number of the wells in the disposal tract were routinely tested, Thornthwaite had his own test of water quality. Because of the rise of the water table over the whole area, one well in the northern part of the tract became artesian. Thornthwaite kept a glass near the well site and delighted in taking visitors there for a drink of the free-flowing and pure water. He always took a drink; however, not all of his more skeptical visitors joined him.

The tremendous success of the Seabrook Farms wastewater disposal operation encouraged Thornthwaite to refine and expand the Seabrook system and also to accept requests from other food processors to help them with the disposal of their effluents. Because he recognized the value of the vegetation cover, he never attempted to dispose of any effluent that might be toxic to the vegetation itself. Originally systems developed for Campbell Soup Company and several others stressed the need for spray irrigation in wooded tracts. Later it was recognized that cultivated tracts could be developed to have high infiltration and percolation rates after a few years of no cultivation, and even hillsides underlain with impermeable shale could be utilized if care was taken to allow water from small irrigation nozzles to flow overland in a thin sheet downhill through a good vegetation cover. Thornthwaite used both honeysuckle and grasses as covers in several systems for Campbell Soup, H. J. Heinz Company, Del Monte Packing Corp., Mrs. Paul's, M and M and others. When heavy snow was a problem, he developed a bubbler system in which the water would be bubbled out on the ground through short risers from buried distribution pipes. The effluent would spread out on the soil surface underneath the snow cover.

This disposal of industrially polluted effluent by means of spray irrigation, which was pioneered by Thornthwaite at Seabrook Farms and is possibly the largest system ever installed for an industry, provided a radically different solution to a complex problem of growing importance to industry. By maintaining and encouraging vegetative growth over the disposal tract and adjusting the volume of effluent disposal to the hydrologic characteristics of the selected area, it was possible not only to maintain the purification potential of the soil and vegetation cover but actually to improve it.

As the years passed, the Seabrook area witnessed the growth of vines on the dead trees, and the whole site took on the look of a tropical rainforest. A change in the species of vegetation occurred, of course, because of the large quantities of water added, and certainly groundwater recharge was greater than that before operations began. However, as long as operations were maintained within the limits dictated by the physical characteristics of the disposal site, the results were satisfactory. If the effluent contained no materials toxic to the vegetation cover, or solids that would clog soil pores and reduce infiltration, Thornthwaite concluded that spray disposal of factory effluent could even be applied to the polluted effluent of industries outside the food processing field. In today's environmental regulatory climate, there would perhaps be need for special permits from state or local agencies for such wastewater disposal, but the use of a poor-growth woodland for the project and the great value of reconditioning such a large volume of effluent suggests that such permits would be forthcoming.

Although a climatologist by occupation, Thornthwaite did not consider wastewater disposal to be outside his field of expertise because he was interested in all aspects of energy and water: above the surface, at the surface, and below the surface of the earth.

Chapter 6

C. W. Thornthwaite Associates

The success that Thornthwaite had at Seabrook Farms from 1946 to 1952 was described in numerous farm publications as well as scientific journals. Thornthwaite's reputation as a consultant in agricultural climatology was growing, and the reputation of the Laboratory of Climatology as a research arm of both the Johns Hopkins University and Seabrook Farms was spreading. To handle this growing consulting activity he established a separate entity, C. W. Thornthwaite Associates, as a partnership with his brother-in-law, Floyd Slentz. The company would handle the outside consulting, utilizing the talents of the workers at the Laboratory of Climatology, Seabrook Farms, and Johns Hopkins. It was conceived as the most effective means of distributing the work activities of the laboratory personnel among contract work for the Department of Defense, consulting work for Seabrook Farms, and outside consulting. Although this distinction became blurred in later years, as the Johns Hopkins Laboratory of Climatology became the Drexel Institute of Technology Laboratory of Climatology and, finally, the C. W. Thornthwaite Associates Laboratory of Climatology, in the beginning it was quite clear whether a person was working for Thornthwaite Associates or the Laboratory of Climatology and how much time was being spent on each activity. Time sheets were filled out daily so that the administrative officials at Johns Hopkins could determine pay status and complete the time activity reports required by government contractors.

Floyd Slentz, known affectionately as Uncle Floyd to all at the laboratory, was a most pleasant addition to the laboratory family. He lived with the Thornthwaites while he was connected with Thornthwaite Associates, and he and Warren were close friends who shared both personal and business interests. Uncle Floyd did not pretend to be a climatologist, or even a researcher, but confined his activities to carrying out Thornthwaite's ideas, writing and answering letters of inquiry, and attempting to establish C. W. Thornthwaite Associates as a private profit-making organization within the Laboratory of Climatology. It was Uncle Floyd who took the lead in trying to commercialize the Cropmeter and to interest seed producers in using growth units to maturity rather than days to maturity in their seed catalogs. As inquiries from individuals and from agricultural enterprises mounted, it was Uncle Floyd who took care of many of the preliminary negotiations and wrote letters on behalf of Thornthwaite to obtain the information needed to develop consulting arrangements. As friend to all, and "father confessor" to some of the younger employees at the laboratory, Uncle Floyd was a beloved and stabilizing influence in the close-knit laboratory staff. More approachable than Thornthwaite himself, he served a very significant role as a buffer, as a sounding board, and as a giver of wise counsel to all.

Originally created to promote the Cropmeter, C. W. Thornthwaite Associates soon became established as a significant consulting group in other aspects of agricultural climatology. One of its earliest attempts was to establish an agricultural weather service with local farmers in the southern New Jersey area. This started with a meeting in early 1954 of Thornthwaite, Slentz, Mather, and a small group of farmers in Hammonton, N.J. Thornthwaite attempted to explain his ideas for an irrigation scheduling service that could be provided to these farmers on an individual basis for a relatively small sum of money. The local farmers would be supplied with rain gages, though most of them already had several of their own at convenient locations in their fields. They would then relay the information on daily rainfall amounts to Thornthwaite Associates, which would combine that information with data on evapotranspiration losses and inform the farmers about when and how much to irrigate their various fields. The service would be tailored to each individual farm because the amount of rainfall varied appreciably from farm to farm. Three of the farmers present at the meeting signed up for the service on a trial basis. Later, one dropped out, but the other two continued for the first year.

The agricultural weather service that Thornthwaite envisioned did not prove to be an unqualified success, in large part because most farmers failed to see irrigation as an integral part of their farming operation, rather than as a last-ditch effort to prevent total crop failure. After dealing with local farmers for two years in trying to operate this weather service, the Thornthwaite Associates personnel concluded that the process was too time consuming and that there were too many other requests for their services. The local agricultural weather service was terminated.

Several fascinating consulting arrangements were established during the early 1950s. The requests for help in disposing of processing plant effluents following the success at Seabrook Farms led Thornthwaite to establish an Effluent Disposal Branch within C. W. Thornthwaite Associates headed by D. W. Parmelee, formerly an irrigation engineer with Seabrook Farms.

Another client was a large tobacco company headquartered in New York. A company officer wrote to Thornthwaite to ask about the possibility of supplemental irrigation on shade-grown tobacco in central Connecticut. This was a new and interesting problem, and Thornthwaite responded immediately, saying that he would like to meet with appropriate corporate officials to discuss the problem. A meeting was soon arranged at a farm site in Connecticut involving Thornthwaite and Mather from Thornthwaite Associates, several corporate officers from New York, and selected field managers from the Connecticut farm division.

Thornthwaite was not particularly familiar with tobacco, considering that he had grown up in central Michigan and spent his college years in Berkeley and Oklahoma. Shade-grown tobacco was even more foreign to him. Much of the early portion of the meeting was spent in understanding the process of raising shade-grown tobacco. This particular tobacco was used as the wrapper for expensive cigars; thus, it had to be perfect in terms of color, taste, and appearance, with no holes or rips in the leaf itself. It was a premium-priced product. Company officials estimated it cost $4,000 an acre to grow the tobacco, but a normal yield would bring some $8,000 an acre. Thus, it would be well worth putting considerable effort into growing better shade-grown tobacco or increasing yields. They felt that scientific irrigation was the answer.

Thornthwaite was fascinated by the complex farming procedure, including the acres of cheesecloth held up by wooden poles. One of his early questions concerned the exact effect of the cheesecloth on the tobacco. In other words, why did they have to grow the tobacco under cheesecloth? This would seem like a routine question to ask, but it quickly showed the real lack of understanding by the tobacco company of the most basic part of its farming operation. Answers came from three tobacco officials. One said it diffused the intense rays of the sun and thus prevented the tobacco leaves from being burned or "overcooked." A second said it served to reduce wind velocity and to protect the leaves from damage by either hail or wind. A third felt that the higher humidity under the cheesecloth kept the leaves bathed in always-moist conditions. Clearly, they were not certain what the cheesecloth did, but they all agreed that it was needed if they were to obtain a high-quality product.

Anders Ångström and Warren Thornthwaite studying the effect of the cheesecloth used in shade-grown tobacco on the receipt of radiation in Seabrook experimental area in 1955. (Photography by E. Taubert, reproduced with kind permission of the Seabrook Farms Co.)

Then Thornthwaite asked them to explain the harvest and curing procedure. It should be noted that these first lines of questions had nothing to do with irrigation problems. This was a typical Thornthwaite approach, for he felt that most individuals or organizations who asked for consulting advice did not really know what their problems were. It was up to the consultant, acting as a diagnostician, to probe and analyze until the real problem could be uncovered and solved.

The harvest procedure was carefully explained: Tobacco leaves grow alternately up the main stalk of the plant. When it was determined that the lower leaves were ready to harvest, the three bottom leaves would be picked and sent to the curing barns where, after a number of months of curing, they could be looked at to determine which were mature. Three leaves were picked to try to ensure that at least one of them would be mature. They were willing to discard two-thirds of their harvest to obtain one mature leaf in each picking.

There was then a brief discussion of irrigation problems, but Thornthwaite felt that irrigation scheduling would be a relatively routine undertaking and would not involve any new concepts or understandings. He was fascinated, however, by the many other growing and harvesting problems faced by the industry. Growers had achieved what they considered to be satisfactory yields with the old methods of farming, and they felt that only scientific irrigation was needed to improve yields. Thornthwaite's report to company officials not only outlined a program of scientific irrigation, but it also contained a detailed proposal to study other aspects of the operation, including the role played by the cheesecloth cover and the harvest procedure. Thornthwaite took his role as a consultant seriously and tried to offer ideas to solve all the problems he saw, not just the one he had been retained to solve.

He proposed to undertake detailed microclimatic observations in the tobacco fields, both with shade and without shade. Because no one seemed to know the particular effect of the shade, he felt it might be possible either to do away with the shade altogether, to replace it with a cheaper product that created the same environmental conditions, or to find a climate someplace in the world that had the desired properties of the shaded fields in Connecticut so that the tobacco could be grown without the expense of installing acres of cheesecloth. Company officials were interested in Thornthwaite's proposal, and they saw some need to understand the ultimate role of the cheesecloth, but they felt that more study of the proposal was needed.

Another proposal involved undertaking a detailed study of the development of the tobacco plant. Thornthwaite was already deeply involved with phenological studies of peas at Seabrook Farms, and it was his idea to use the plant as an indicator of how development was progressing. He reasoned that similar phenological studies of tobacco might reveal important information on the rate of progress to maturity of individual leaves on the plant. Because tobacco leaves come out around the stalk as the plant grows, he felt that there might be a good relation between what happens near the growing top of the plant and how the lower leaves mature. For example, when the twentieth leaf on the stalk begins to emerge, the first leaf at the bottom of the plant might be mature; or, when flowers develop at the top of the plant, the tenth leaf might be ready for harvesting. This proposal called for a simple phenological study of the plant and gave considerable promise to improve the yield of mature tobacco leaves. Company officials agreed to have Thornthwaite Associates undertake that study, along with the irrigation scheduling program.

It was unfortunate for the research on shade-grown tobacco that rapid changes were then taking place in the industry: a new product, homogenized tobacco, was being developed. It was possible to take a lower-grade tobacco leaf (even one with holes or rips), add materials to it to create the color and taste desired, process the tobacco much like wood pulp, and then roll out a whole sheet of tobacco as one would paper. In that way, the cigar producers could obtain just the type of tobacco they wanted for their wrappers. The increased use of this homogenized tobacco very quickly put to an end the need for shade-grown tobacco and the concern about how to pick leaves at the peak of maturity. Thornthwaite Associates began the phenological studies on tobacco but was never able to bring the study to a successful conclusion because of the rapid shift in the industry to "manufactured" tobacco leaves.

A second consulting activity in agricultural climatology involved the scheduling of supplemental furrow irrigation on a large cotton plantation in Arkansas. Thornthwaite had never applied his irrigation program to furrow irrigation, and as before, he saw the opportunity as one of learning for himself as well as improving cotton production through a more scientific agriculture. Although the Arkansas farmland was reasonably level to start with, it was still necessary to bring in giant scrapers and bulldozers to create a gradual field slope, backhoes to create the canals to carry water to the borders of the fields, and well drillers to establish the necessary wells to supply adequate water. No cotton could be grown until the land renovation work was completed.

In the meantime, Thornthwaite undertook his own preliminary research on the problem. He had no doubt that he would be able to provide a practical and workable irrigation program; however, he also wondered whether irrigation was the only agricultural problem that limited yield. Thus, as was his custom, he studied the literature to see what could be learned about cotton production, and he also began to study the plant in the field and in his own garden, even though he realized that southern New Jersey was too far north to obtain much of a yield of cotton. Phenological studies of cotton in Arkansas, along with basic climatic research, provided information on the number of growth units that would be accumulated in a normal growing season. Farmers in the area were satisfied if they were able to harvest eight mature bolls from each plant annually, which provided a yield of one and one-half bales of cotton to the acre and a reasonable margin of profit. Thornthwaite found it most disturbing that his studies showed there was enough energy and time in the Arkansas area to produce eighty mature cotton bolls per plant. He wondered how the farmers could be satisfied with just one-tenth of the possible yield and, more importantly, how their farming habits eliminated ninety percent of their potential yield.

Study of the development of the cotton plant and the practices employed by the cotton growers in bringing the crop to maturity suggested to Thornthwaite a number of ways in which yields were being limited by growing practices. For example, the need for irrigation for cotton was determined by looking across a field of growing cotton into the rising or setting sun—if the field had a blue tinge to it, this indicated that the plants were suffering from a lack of water and needed irrigation. Thornthwaite pointed out the fallacy of this thinking by stating that irrigation was needed before, not after, the color change was noticed. After the color had changed, indicating plant stress, some yield had already been lost. "Would you want help before or after you turned blue?" he remarked facetiously, but his point was driven home.

Other practices also served to limit yields. "Insect scouts" (agricultural students from the University of Arkansas) were employed by the cotton farmers in the state to survey fields for weevils and other harmful insects. They were trained to call for insecticides if the infestation of weevils was greater than a certain density. Thornthwaite did not like this technique, for he felt this practice allowed too great a reduction in yield before insecticides were applied. He spoke to the entomology professors at the university and accused them of turning out "bale-and-a-half thinkers." By this he meant that their program, taught to insect scouts and others in the agricultural college, was predicated on obtaining yields of one and one-half bales per acre. If that could be achieved, the farming process was considered successful. Thornthwaite argued that if they would modify their programs—call for insecticide application before weevils reached the critical limit or irrigate sooner—they might achieve a yield of more than a bale and a half.

Thornthwaite realized that other activities that limited yields included the manner in which weeds were eradicated. Geese were often employed to take care of small weeds, and although they were not very effective in large fields, they probably did not do much damage to the cotton. Small jets of fire were used to burn the weeds, but the equipment often caused mechanical damage to the cotton plants. Probably the greatest damage came from the regular use of cultivators in the fields, for as the cotton plants became larger, there was little room for the wheels of the tractors and the mass of the cultivators to pass between rows of plants without hitting them and possibly knocking off developing bolls. Lack of sufficient fertilizer probably also reduced yields, especially if the cotton was being irrigated. As more water was supplied, more fertilizer would be needed to maintain the optimum amount needed for maximum growth.

The various studies that Thornthwaite wanted to undertake in Arkansas were delayed because of the need to allow the soil profile to reestablish itself after the land-leveling operations, and also because of two years of adverse weather conditions. One very cloudy year brought a severe outbreak of mildew and fungus that eliminated any possibility of significant yields, and this was followed by a year with so much rain in the harvest season that it was impossible to deploy any harvesting equipment. This succession of unfortunate developments convinced the farm owner to sell out and move to Arizona, where such weather-related problems were less likely. Thus, Thornthwaite's work on cotton irrigation was terminated. Although unsatisfied, he recognized that he had not failed. He did regret not having had the opportunity to see whether his proposed program could have produced more than eight mature bolls per plant!

From these examples, it can be seen that Warren Thornthwaite was eclectic in his scientific interests. He was trained as a geographer and undertook early work on the urban geography of Louisville, Ky., the analysis of census data, problems of climatic classification, the moisture factor in climate, and regional studies of the Great Plains. As his interest in evaporation and microclimatology increased over the years, he recognized the need to obtain better observations of the climatic factors of energy and moisture near the earth's surface. This was exemplified by his year-long study of evapotranspiration from a grassed surface in Arlington, Va., near the site of the present Pentagon building. During this study, he had the available anemometers and psychrometers modified and made more responsive to rapid changes in wind and moisture conditions. This allowed him to determine small additions of moisture to the air at several levels as it moved horizontally over a flat, vegetated area. However, the process also made him realize that slow-response instruments severely limited observations, and this initiated a lifelong passion to develop improved meteorological sensors.

Thornthwaite was not trained as an electrical engineer and had no real background in instrument design, but he had an innate understanding of how instruments should work in order to provide the type of observations that he wanted. In later years, when sophisticated instrumentation was being designed and built in the shops of the Laboratory of Climatology, Thornthwaite was always deeply involved in the design steps. He knew what the instrument needed to do, and he could describe how it had to be designed in order to achieve that end. Although he left the actual fabrication of the instruments to others, he continually looked over their shoulders and followed the process step by step. In the course of the instrument development work, five patents were issued to Thornthwaite for various improvements and modifications to instruments he helped to design. Unfortunately, three of these patents were not issued until after his death, but they bear witness to his genius in instrumentation.

Patent No. 2,240,082 was issued to C. W. Thornthwaite on April 29, 1941, for new and useful improvements to an apparatus that indicated changes in the moisture concentration of a gas over time. This improved ability to measure small changes in the dew point of the air over short time periods went a long way in solving the problems that had arisen during his year-long study of evapotranspiration from a grassed field—or from any field of vegetation. A second patent (No. 2,268,785), issued on January 6, 1942, covered further modifications to the dew point recorder and indicator identified in the earlier patent.

In 1955, the instrument department of the laboratory completed the development and production of a new wind-profile register system for precise micrometeorological work. The anemometers that formed a part of this system were extremely sensitive to minor fluctuations in wind speed and direction. The register employed a two-stage transistor amplifier circuit. The system was portable and could be operated from a six-volt battery and thus was not dependent on the availability of electric power. The register had already been selected by the Quartermaster Corps, the Arctic Institute of North America, and a university group for use in their own research programs. Other development work led to the production of an improved dew point hygrometer for microclimatic work, radiation shields for use in temperature measurements, and a new system for measuring soil temperatures without disturbing the soil profile.

During 1956, methods for matching anemometers were perfected, and special matched sets of anemometers were built for eight research groups. Four of these sets were put into service in the Arctic (one on a drifting ice station in the Arctic Ocean) to obtain observations for the International Geophysical Year of 1957–1958. In the development of the dew point hygrometer, a new method of cooling the mirror utilizing thermoelectric principles was explored.

During 1958, the laboratory concentrated primarily on modifying the dew point hygrometer so that it would be capable of giving values of dew point at air temperatures less than −40° or −50°F. A new method of cooling the mirror surface using canned Freon gas was employed with great success. In the same year, the laboratory placed major emphasis on the fabrication and field testing of a propeller-type instrument to measure the vertical motion of the wind, as well as the construction of a lightweight, portable mast on which to support meteorological instruments. The mast was one hundred feet in height, weighed only thirty-seven pounds, and was made of five twenty-foot sections of two-inch aluminum irrigation pipe. The tower was assembled horizontally, raised to the vertical position in one piece, and was supported by four sets of guy wires. The raising and lowering operation required only a few minutes' time, and the mast could be transported easily in sections on the top of a car or light truck.

Patent No. 3,166,928, issued January 26, 1965, a year and a half after Warren's death, was in the name of Thornthwaite, Justin Jackson, K. Ray Ono and W. J. Superior, of whom the latter three worked in the instrument shop of the laboratory. The application for the patent was filed June 28, 1961, and covered an improved apparatus for measuring and indicating the moisture concentration of a gas. This involved the development of "a more efficient apparatus for bringing a light-reflector to and maintaining it at the dew point of a gas sample in contact with the reflector" and "to provide ways and means for automatically heating and cooling the reflector, as conditions demand, irrespective of the magnitude and sign of the difference between the reflector temperature and the dew point of the gas sample" (U.S. Patent Office, 3,166,928, patented January 26, 1965 p. 1). The new device provided instantaneous and continuous readings of the dew point of a flow of gas past the reflector surface, always compensating for slight fluctuations in the intensity of light from the light source and in the temperature of the ambient air to the heat pump.

Patent No. 3,208,275, issued September 28, 1965 (more than two years after Thornthwaite's death), involved an apparatus to measure vertical wind and climatological fluxes. The patent was issued to Thornthwaite and W. J. Superior. In the patent application it is pointed out,

> While horizontal wind may vary in azimuth from instant to instant, it rarely, if ever, suffers the complete reversal of direction that is a characteristic feature of vertical flows. The vertical flow may change from "up" to "down" at very short intervals, such as a fraction of a second, and these changes greatly affect the time-average or net vertical flow; even though the flow in the respective directions persists for only such short intervals.
>
> It is accordingly a principal object of this invention to provide methods and apparatus for the precise registration or recording of the vertical components of air flow, and of any measurable quantities associated with such flow. (U.S. Patent Office, 3,208,275, patented September 28, 1965, p. 1)

The final patent (No. 3,241,436) was issued March 22, 1966, nearly three years after Thornthwaite's death, and covered a direct digital printout data recording apparatus. Patent holders were Thornthwaite, W. J. Superior, and K. Ray Ono. The application, filed just two months before Thornthwaite's death, stated,

> The present invention aims to provide a compact, self-contained apparatus including a plurality of individual numerical registers together with equipment for photographing their readings or indications in serial order, at predetermined periods, on a single stationary frame or photoprint area, to yield a composite photographic record covering a predetermined period of time. At stated times, the user can readily remove the fully processed digital record for the preceding interval, and advance a fresh film or print section for recording

in the next interval. (U.S. Patent Office, 3,241,436, patented March 22, 1966, p. 1)

The photographic record was obtained by using a Polaroid camera connected to the digital registers and a timing device to maintain the predetermined period of registration.

Other devices and equipment were developed by Thornthwaite and his colleagues at the laboratory. Some were patentable, some were not, but given that they were rather specialized and had limited marketability, there was little reason to press for additional patent protection.

Although C. W. Thornthwaite Associates had originally been conceived as a partnership between Thornthwaite and his brother-in-law, Floyd Slentz, to sell the Cropmeter, it turned out to be a most serendipitous organization. It provided the framework for bringing together one of the first groups of consulting climatologists, and it demonstrated that such a group had the ability to solve practical climatic problems in agriculture and industry. No single individual involved in the organization had the skills necessary to solve every problem that might arise, but the combination of individuals with differing talents provided the opportunity to achieve successful solutions to a wide range of problems. Thornthwaite Associates might not ultimately have been the best possible vehicle for such a consulting group, but it functioned well at the time under the strong centralized leadership of Warren Thornthwaite. Today it still provides a viable model for future consulting climatologists to consider.

Chapter 7

Thornthwaite and Academic Geography

Although he had a degree in geography and had taught geography courses at the University of Oklahoma, Thornthwaite waxed hot and cold about academic geography. All his academic associations were with geography departments, but he never liked to submit himself to the rigors of the academic year, including regular classes. After he moved to Washington in 1935, he spent nearly a decade without an academic home while he was developing his research work with the Climatic and Physiographic Division of the Soil Conservation Service.

In 1945 O. E. Baker, whom Thornthwaite had known in the Department of Agriculture and who had accepted the position of head of geography at the University of Maryland in College Park, invited Thornthwaite to teach a course in climatology at that university. Thornthwaite was living in College Park at the time, so he accepted the invitation and agreed to give an evening course for graduate students. The number of students was small, usually three or four, and this suited Thornthwaite because he liked small classes and wanted to try out his ideas on the water balance for his new climatic classification on the graduate students. One of these students was Marie Lustig (later Sanderson), who fell under the "spell of the water balance" and became Thornthwaite's first graduate student. She wrote her master's thesis under his direction and then returned to

Canada in 1946 to carry out some of Thornthwaite's field research at the Ontario Research Foundation in Toronto.

Thornthwaite remained at the University of Maryland for only a short period, and in 1948 he became a professor of Geography and Agricultural Climatology at the Isaiah Bowman School of Geography at the Johns Hopkins University. This association was acutally established so that Johns Hopkins could manage the financial aspects of the contract that had just been negotiated among Seabrook Farms, the Air Weather Service, and Thornthwaite. He did offer a graduate course in climatology at Johns Hopkins, but about halfway through one semester he tired of traveling to Baltimore from Seabrook and let the course terminate.

He was busy building up the Laboratory of Climatology at Seabrook at the time, and he wanted the laboratory to be associated with a first-rate university, as he had clearly stated in his 1943 article "Status and Prospects of Climatology." He was not quite satisfied that the laboratory, associated with Johns Hopkins, was about a two-hour drive from the main campus. He preferred that the students come to him in his laboratory.

Thornthwaite had never really liked teaching in a formal classroom setting; he preferred a mentor-pupil relationship in the field. He had a rather slow, pedantic teaching style. He moved slowly when standing in the classroom and often put one hand on a desk or table, probably to steady himself, given that he suffered an inner-ear disease that caused a slight loss of balance with motion. He selected his words carefully, often changing words or giving several choices of words in trying to make a point. He seemed uncomfortable with the formal lecture style, which made an unfavorable impression on those students who were not excited by climatology. Thornthwaite preferred to work with dedicated climatologists who were eager to learn and could overlook his teaching style. He was most in his element in the field, joining in an experiment, making a set of measurements, or offering instructive comments on what was happening or why the results were coming out the way they were. In these informal and relaxed situations, he was a most effective teacher.

The laboratory had a fine library in climatology, better than that at Johns Hopkins, and the measurements of wind, temperature, and humidity fluxes; of evapotranspiration; and of solar and net radiation were all carried out at Seabrook. Thus, he saw no reason why his climatology students should have to go to Baltimore and take time to learn other aspects of geography when all the climatology instruction they needed was available to them at the laboratory. Thornthwaite began to work for a separate program in climatology at Seabrook in the summer of 1950, when he announced the availability of three courses through the Isaiah Bowman School of Geography: microclimatology, taught by Rudolf Geiger; climatology; taught by C. W. Thornthwaite; and a climatology field course with Thornthwaite, Geiger, and staff.

The inauguration of such courses was what Thornthwaite had hoped to achieve, and with such distinguished visiting experts as Rudolf Geiger and others, he had visions of a strong academic program in climatology at the Laboratory of Climatology. However, the fact that the five original students in climatology were registered for administrative purposes in the Bowman School of Geography presented a problem. One student was an undergraduate and soon had his academic program interrupted by military service, and a second dropped out to return to a geography program. The three remaining graduate students—Russ Mather, Maury Halstead, and Donald Portman—all spent varying times in course work in geography at the Baltimore campus as well as in course work and practical field training at the laboratory at Seabrook Farms.

Thornthwaite increasingly felt that his climatology students should not have to be subjected to the wide range of geography courses offered at the Bowman School but, rather, that they should concentrate on the climatology, statistics, and physics necessary to become first-class climatologists, with the bulk of the work being done at the Laboratory of Climatology. As will be seen in later correspondence between George Carter and Thornthwaite, the crux of the problem was that Thornthwaite wanted to turn out first-rate climatologists who were also well trained in mathematics, statistics, physics, biology, and other science-oriented subjects. On the other hand, Carter knew that, under university regulations, the climatologists' degrees would be in geography, and therefore he felt that the students should have a minimum understanding of other aspects of geography. Although Thornthwaite understood the problem, he was not willing to compromise his belief that his students should be trained as climatologists rather than as geographers. In fact, as stated in his presidential address to the Association of American Geographers, "The Task Ahead," in 1961, he felt that geographers should move toward the rigorous, science-oriented training he prescribed for climatologists.

Thornthwaite's turning away from the traditional course work in geography can be seen in his January 19, 1951, letter to Lieutenant Colonel Arakelian, of the Air Weather Service headquarters, who had asked for information on possible training available at Johns Hopkins in meteorology and other geophysical subjects: "Instruction in climatology is carried on partly on the main campus of the University and partly at the Laboratory. The graduate students divide their time between the two locations. At the University the students take formal courses mainly in other departments; i.e., physics, mathematics, statistics, aerodynamics, etc. At the Laboratory they also take formal courses given by members of the resident staff and also by professors who come over from the main campus. Principally, however, they are engaged in field work and in independent research."

Both Mather and Halstead had their dissertation defenses in the spring of 1951, and although both passed, it was apparent that some of the faculty of the Bowman School were concerned about giving geography degrees to climatologists who had a minimal geography background. The growing insistence by Thornthwaite that he direct the curriculum for each of his students, and that it need not include those aspects of geography that he considered irrelevant for climatologists, raised significant questions.

George Carter, chair of the Bowman School, wrote his staff on June 1, 1951, pointing out that "a critical decision must be made concerning work in climatology." He wrote,

> Men may come for technical training in micrometeorology, but will receive no degrees. Men may take degrees in geography, at the Bowman School specializing in climatology. Men can take degrees in climatology or bioclimatology through the Physical Science Group or the Biological Science Group. They will not be required to meet *geographic* requirements.
>
> Implicit in these statements is the feeling that there is not sufficient range of interests and fields represented in so limited an environment as Seabrook to compensate to any sizable degree for the vastly more varied fare offered at Homewood [the main Johns Hopkins campus]. We recognize the Seabrook Laboratory as a unique field laboratory of great value for full-time residence to students advanced to the thesis level, and of equal value to beginning students for shorter periods (summers especially) to give practical direction and meaning to their further studies at Homewood. (personal communication from G. Carter, June 1951)

There were several faculty meetings to discuss the problems recognized by Carter, as well as his proposed recommendations, but no decision was reached. Both Thornthwaite and Carter remained adamant in their views about the need for some breadth in geography for students in the climatology program. With several more doctoral students in climatology working their way through the system, Carter felt that it was time to make a firm decision, and so he wrote to President Bronk of the university (as well as to Thornthwaite) on November 19, 1951, recommending the separation, both administratively and academically, of the Laboratory of Climatology from the Isaiah Bowman School of Geography. He suggested that the Laboratory of Climatology be set up as an affiliate or institute of the university, similar to the university's arrangement with the Chesapeake Bay Institute. Carter concluded,

> Dr. Thornthwaite has characterized the work demanded of his men in the Bowman School as "not useful for my men" and as a "waste of their time." He finds statistics as taught either on the Homewood or Medical Campus non-useful also. This is a general complaint concerning basic courses as preparation for specific work. However, I do not think it valid, and consider the alternative—Thornthwaite's students teaching Thornthwaite's students in the isolation of Seabrook—quite unacceptable for producing geographers.
> Thornthwaite has a contribution to make that is peculiarly his own. I wish to set him free to do his work his way. The opposition of the staff of the Bowman School to such course work at Seabrook has been repeatedly and forcefully expressed to Dr. Thornthwaite. (personal communication from G. Carter, November, 19, 1951)

Thornthwaite seemed surprised at this action by Carter, although it is somewhat difficult to understand why he seemed unaware that his increasing unwillingness to cooperate with the faculty at the Bowman School was leading to the possible separation of the two activities. He wrote the following to Carter on November 21 defending his actions:

> On receiving the copy this morning of your memorandum to President Bronk, and noting that it must have been written immediately after I left Rogers House [the Bowman School] for Seabrook, I wondered why the matter was not included for discussion in the staff meeting or why you did not so much as hint of your plan to me. I must say I honestly don't understand the difficulty. . . . the Laboratory, as it has developed, is an asset to the Bowman School; our research, however detailed, fits readily into the broad field of geogra-

phy; our training facilities in no way displace those at Homewood, but supplement and amplify them.

How can you say that our work is non-geographic? Climatology is one of the mainsprings of geography. As a result of a paper on climatology that I published in the *Geographic Review* in 1948, I was made an Honorary Member of the American Geographic Society and a Corresponding Editor of the *Review*. One cannot do worthwhile work in climatology without physics and mathematics. Our study of atmospheric turbulence is not an end in itself, but to determine evaporation from natural surfaces so as to evaluate the moisture factor and delimit the moisture regions of the earth, thus to arrive at their agricultural possibilities and finally their human potentials. . . .

Our chief difference seems to be in our concepts of the proper function of the Bowman School. In my view, the Bowman School is set up to develop competence in any of the important branches of geography. Climatology is one such branch. My academic title is Professor of Climatology. The Laboratory of Climatology was set up in the Bowman School to train competent climatologists. We might look forward to a development of cartography in similar fashion. Obviously, no one could deal competently with map projections without a solid mathematical foundation.

You are well acquainted with the deficiency of geography in this country. Geographers get a smattering of map projections, of climatology, of soils, of vegetation, and so on; always only a smattering because the student is not capable of getting more and the professor is not competent to give more. As you have yourself said, it is all a fraud. Mr. Bowman recognized this too, and that was what he was trying to avoid in setting up the Department in Hopkins.

We should not be trying to produce geographers in the sense that other Universities are doing. Such training cannot avoid producing superficial students who end in frustration when they attempt a real job out in the world. The only things that such students can do is teach others in the same sterile discipline. The alternative is to give our students real competence in a branch of geography where their interests lie. I don't deny that you can do that in your field. I am trying to do that in climatology, but with the help of others from all over the world. When you "consider the alternative—Thornthwaite's students teaching Thornthwaite's students in the isolation of Seabrook—quite unacceptable for producing geographers" you are talking nonsense. My students are climatologists. They are not rounded geographers because no one ever can be. Actually, as you know, we do not preserve the fiction that any of our students are well rounded. We have no work in cartography and practically no regional courses.

My students are in great demand as climatologists and but for the vision that I have given them I probably should not be able to keep them. The Bowman School is fortunate in having my students teaching climatology. For the first time there is in the United States a place where a student can be trained in climatology. Certainly that is a good thing for geography and a good thing for the Bowman School.

The association with the Bowman School essentially ended at that time. A mimeographed document dated March 28, 1952, outlined "Graduate Instruction in Climatology" at the Johns Hopkins University, Laboratory of Climatology, Seabrook, N.J.:

> The student will be expected to arrive in June and to spend the first summer at work on the practical problems of climatology at the Laboratory. This will be a probationary period for all students and permission to continue in the academic work will be based on the individual's demonstration of ability.
>
> The basic requirements for acceptance of students . . . are essentially those listed by the University as "Requirements for the A.B. degree in the Physical Sciences Group."
>
> Upon successful completion of the probationary period of summer work at Seabrook the student will spend at least one academic year at the University in Baltimore obtaining the required courses of instruction. In addition, a minimum of one more summer and one full academic year will be spent at the Laboratory at Seabrook completing the climatological requirements. A total of two years of residence will be required for candidacy for the Ph.D. degree. (Unpublished memo from Thornthwaite to prospective students, March 1952)

Thornthwaite, cut off from the Bowman School connection, still pursued his dream for a climatic institute to provide significant, in-depth training in climatology to specially selected students through the Physical Sciences Group of the university.

By November 1953, the university had decided either to bring the laboratory back to Baltimore—and thus bring it more under the direct control of the university—or to terminate its association with the laboratory entirely. Thornthwaite did not want to move, and he also recognized the good research opportunities available in the flat, open fields of the Seabrook area. He rejected the move to Baltimore, and the laboratory began its second period of existence while a new academic sponsor for its research activities was sought. Thornthwaite had tried geography and been rejected; he had tried running an academic program through the Physical Sciences Group and been unsuccessful. He wanted the laboratory to continue, but he was so busy with the research and consulting at the laboratory that he had to postpone his desire to develop an academic program there, and it actually never resurfaced during the remaining nine years of his life.

The criteria for a new administrative sponsor were dictated by the Department of Defense contracts that were then being undertaken by the laboratory. The government wanted the financial aspects of the contract work to be handled by a nonprofit (preferably academic) institution. Thornthwaite therefore set out to find another nearby academic institution that would be willing to sponsor the laboratory and perhaps allow it to continue academic course work. His first set of inquiries were directed toward the University of Delaware, which was just forty miles from the laboratory's new location in Centerton, N.J. This move had been necessitated by the termination of Thornthwaite's consulting arrangements with Seabrook Farms and its need to reclaim the laboratory building for an expansion of its own operations. Thus, in the fall of 1953 and the spring of 1954, Thornthwaite was faced with the dual problems of locating new quarters and a new sponsor. Both problems were solved in the spring of 1954, when he personally bought two buildings in Centerton, N.J., about eight miles from Seabrook. One was a private house and the other a cinderblock building across the street that had been used as a post office for Centerton and as a barber shop. Thornthwaite and his secretary, June Yoshioka, moved into the cinderblock building while the rest of the laboratory personnel and the extensive library moved into the two-story wood-frame house across the street. Part of the library was housed in the basement and the rest on the walls of many of the offices in the house.

Thornthwaite visited various administrative officials at the University of Delaware, including President J. A. Perkins and Dean George Worrilow of the College of Agricultural Science. Thornthwaite felt that a more useful association at the university would be through agriculture rather than geography. Actually, at that time, there was no Department of Geography at the university, and the few courses in geography that were offered were in a combined geology and geography program. The University of Delaware officials expressed only lukewarm interest in any association with either the Laboratory of Climatology or with Thornthwaite himself. It was clear that they had discussed the problems that had surrounded the association with Johns Hopkins, and they were not certain that they wanted the problems inherent in having an academic unit located in another state at some distance from the main campus. They finally rejected Thornthwaite's overtures for association.

Thornthwaite's next point of contact was the Drexel Institute of Technology in Philadelphia, also about forty miles from the Laboratory. The laboratory had been utilizing Drexel students for several years during the six-month period they had to spend working in industry each year. In the interest of their students, the Drexel Institute officials were much more responsive to Thornthwaite's suggestions than those at the University of Delaware had been. Although they wanted no academic programs through the Laboratory of Climatology, they were more than willing to handle the financial aspects of the government contracts (and, of course, to accept the overhead payments that accompanied the contracts). Thornthwaite was made an adjunct professor at Drexel, even though he was not expected to offer any course work. A new chapter in the life of the laboratory began in 1954, with the Drexel Institute of Technology Laboratory of Climatology located in Centerton, N.J. However, in July and August 1959, talks were held with administrators from the Drexel Institute of Technology concerning a continuing relation between the laboratory and Drexel, and it was decided that the most acceptable future relationship would be for Drexel to subcontract its research work in climatology to the laboratory. Thus, on September 1, 1959, the previous affiliation of the laboratory with Drexel was discontinued and a more clearly defined subcontractor relationship was established.

In 1956, officials of the University of Chicago approached the laboratory to explore the possibility of using the laboratory facilities to train their graduate students in meteorology and geography. These discussions resulted in a tentative program of academic cooperation that allowed the laboratory to offer formal course work leading to graduate degrees in climatology. Beginning in 1957, Thornthwaite spent a week in Chicago during each of the three quarters, lecturing and taking part in the academic program.

A final aspect of Thornthwaite's association with academic geography must be recorded. He possessed a sly sense of humor, and sometimes the object of his humorous stories were geographers. It was not always possible to determine whether the intent of his humor was purely to amuse or to sugarcoat a point he wished to make in a way that might not offend too deeply. Such was the case in an article which he wrote in 1939—but which was not published until 1961, in *The Professional Geographer*—entitled "American Geographers: A Critical Evaluation."

One might ask why he waited so long before submitting the article for publication. In 1939, when Thornthwaite wrote it, he was working for the government. He had already completed a seven-year stretch in the Geography Department at the University of Oklahoma and five years' service with the Department of Agriculture. He was not greatly enamored with geography as an academic discipline and had questions about the training of its practitioners. By 1961, when the article was submitted for publication, he had been selected as honorary president of the Association of American Geographers and had also been honored by the American Geographical Society. He was now more supportive of the field as an academic discipline than he had been in 1939. One might hypothesize that when the article was written it was meant as a veiled criticism of the geography of the day, whereas when it was released in 1961, Thornthwaite saw it more as a humorous spoof of colleagues of an earlier day. It was so interpreted by most of those who read it in published form.

Part of the article is reprinted below with the permission of the Association of American Geographers.

A recent critical survey of current geographic thought is of unusual interest to all American geographers because it supplies a basis not only for a general evaluation of American geographers but for definite ranking of individual geographers as well. There has, in the past, been great uncertainty as to the relative worth of the various members of the Association and consequently a scientific method which can be depended upon to remove this uncertainty is extremely welcome. The method herewith presented is based on the theorem: "By their works ye shall know them."

The works of the American geographers have been critically surveyed. From the index of authors the number of citations in the text can be determined. This number is taken as a rating index on the assumption that the more important geographically a man's contribution is, the more numerous will be the references to his work. Conversely, if a man is not mentioned at all it indicates either that he is not a geographer or that his work is worthless.

The ratings of individual geographers, then, supply an invaluable basis for rating geography departments in the various universities and geographers in the government and elsewhere. . . . A number of interesting things are shown by this analysis. Of great significance is the pattern of distribution of grades of the staff members of the various departments [Definition of grades: 41–50, Genius; 31–40, Superior; 21–30, Good; 11–20, Mediocre; 1–10, Moron; 0, Idiot]. The distribution pattern indicates that some departments are reasonably homogeneous. For instance, at Chicago all members are mediocre and at Clark half are low morons and half are idiots. This is in marked contrast to conditions in Minnesota where two morons and one low mediocre are in the same department with a genius. In Michigan and Wisconsin the situation is similar but not so extreme. In California there is a wide range with two good, and one genius counterbalancing one low moron. . . .

A serious defect in departmental organization at a number of universities is revealed. At Chicago, Clark, Louisiana, Michigan, and Minnesota the chairmen are the lowest rank of the staff members of the geography departments. Most conspicuous is the situation at Minnesota, where a man of grade 45 is supervised by one of grade 3. It might be pointed out in passing that of the three members who are or have been university presidents, two are morons and one is an idiot. It is probable that these anomalies have developed because there has never been in the past a reliable rating of geographers available. It is to be expected that now a redistribution will occur and those qualified to lead will lead.

The fact that the geographers in the government service consist solely of morons and idiots with a preponderance of the latter, explains why no work of any value can come from government departments, and also illustrates a form of natural selection which has been discussed in a presidential address before the Association. The fact that there are still a few morons and idiots in university departments indicates that the process of selection has not yet run to completion. It is to be expected that eventually all of these will be in the government service. . . .

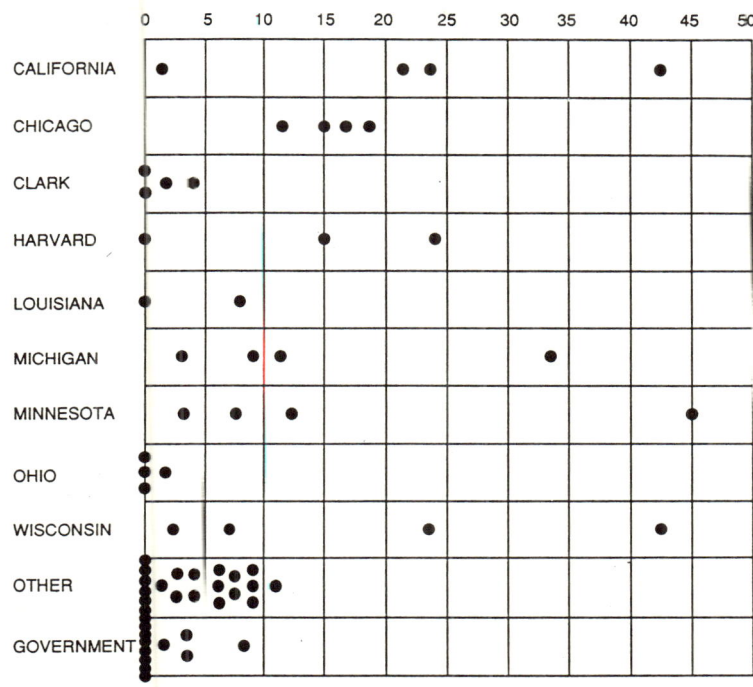

Distribution of Thornthwaite-rated geographers in American universities and federal employment.

In conclusion, the author wishes to emphasize that because of their great value the ratings included in this paper should be disseminated as widely as possible. Particularly should they be placed in the hands of prospective geography students who, through ignorance or through malicious propaganda, might otherwise make a mistake in the choice of school and professor. ("American Geographers: A Critical Evaluation," 1961, © The Association of American Geographers, quoted with permission, pp. 10–12).

Sometimes Thornthwaite's humor took the form of practical jokes. He had developed a close relationship with the *Geographical Review*, which published many of his climate articles, and its editor, Wilma Fairchild. When on trips, he would usually send Fairchild a postcard to keep her informed as to where he was. On one occasion, she remarked to him that he seemed to be traveling extensively for she was getting postcards from different places almost simultaneously. This comment provided the germ of an idea, and he decided to play a trick on her. He wrote postcards and enclosed them in letters to the postmasters in each of the U.S. state capitals with the request that they mail the postcard on the day specified in his letter. He had assigned to one of his daughters the task of determining how long it would take a postcard to travel by mail from each state capital to New York City. Almost all of the postmasters complied, and on the selected day cards postmarked in more than half of the U.S. state capitals arrived on Fairchild's desk. Most of the rest arrived the following day. They all indicated that Thornthwaite had been there and that he had mailed her the card. It was a mystery to Fairchild for some time until Thornthwaite, with a twinkle in his eye, explained how he had pulled off the hoax.

Chapter 8

International Activities:
1947-1958

The American Geophysical Union had a committee on climatology of which Thornthwaite was chair in 1947–1948. The other members of the committee were Erwin Biel, Phil Church, W. C. Jacobs, Helmut Landsberg, John Leighly, and Katherine Hafstad. The 1949 "Report of the Committee on Climatology" is interesting in revealing Thornthwaite's special interests:

My own primary interest continues to be with the measurement of evaporation from natural surfaces. Since the publication of Thornthwaite and Holzman, "Measurement of evaporation from land and water surfaces" in 1942, we have come to realize that success in using the vapor transport method to measure evaporation must await improved instruments for measuring the temperature, wind, and atmospheric moisture and requires revision of turbulence theory. While work is continuing on the improvement of instruments and theory I have sought means of obtaining an approximate empirical answer. In 1946 an evapotranspirometer of my design . . . was put into operation near Mexico City. In 1947 a battery of four evapotranspirometers was established in Toronto, Ontario and another in Seabrook, New Jersey. In 1948 a battery of two began operation in Kapuskasing, Ontario, and the number in Seabrook was increased to six. The evapotranspirometer does not provide the elegant solution to the problem of measuring evaporation that the vapor transport method does, but it is an instrument that is relatively inexpensive and that can be operated by usual experiment station personnel, and I am in hopes that installations will be made in various parts of the world. Wide interest has been aroused in the method, and plans have been formulated to establish stations in Argentina, Israel, Saskatchewan, and on the lower Mackenzie River in the Northwest Territories of Canada. [The Toronto, Kapuskasing, and Northwest Territories experiments were carried out by Marie Sanderson.] I would like to see several stations established in the United States. To obtain a first approximation of the magnitude of evapotranspiration in the United States, I brought together as many observations as I could find on evaporation and transpiration from land areas in the United States (principally irrigation projects in the West and watersheds in the East) and have produced a provisional equation for determining potential evapotranspiration from standard climatological observations. Through use of this formula, I produced a detailed map of potential evapotranspiration in the United States. Much more work remains to be done on this problem and I urge a concerted attack. ("Report of the Committee on Climatology, 1947–48," Transactions of the American Geophysical Union, p. 440)

By 1951, the Laboratory of Climatology already had a worldwide reputation. Thornthwaite's 1948 paper, "An Approach Toward a Rational Classification of Climate," had been available to researchers around the world for three years, and a large number of studies applying the new classification to an evaluation of climates in various countries were appearing. The number of international visitors coming to the laboratory to learn more of Thornthwaite's research was impressive and growing almost daily. Although at that time Thornthwaite had had only limited foreign travel, mainly to Mexico, he was well known to climatologists around the world.

The World Meteorological Organization (WMO) was established in 1947 when it took over the functions of the long-established International Meteorological Organization (IMO). Whereas IMO was an association of meteorological services, WMO was an association of seventy-nine member states. It was created as a specialized agency of the United Nations and was therefore comparable in structure to the Food and Agricultural Organization, World Health Organization, or United Nations Educational, Scientific, and Cultural Organization (UNESCO).

It established close working relations with these other United Nations agencies. WMO had the task of regulating world meteorology because, of all the sciences, it represented the one most dependent on international accord.

The first congress of WMO was held in Paris in 1951, and eight Technical commissions were established to help keep WMO abreast of the problems in each major part of the field. One of these commissions was the Commission for Climatology (CCL). Thornthwaite was not present at that congress because he was not a representative of any member government. In spite of this, he was elected first president of CCL by its members. In his presidential address to the First Session of the Commission for Climatology in Washington in March 1953, he referred to this rather unusual election: "I regard my election to the Presidency of this commission, in Paris two years ago without my knowledge or consent, as further evidence of a change in emphasis. I am not a meteorologist in the narrow sense, but a geographer turned climatologist. I am not now and never have been associated with an official meteorological service except in a consulting capacity. Therefore, my election seems to me to have been a further indication that our commission has a mandate from WMO to do whatever can be done to promote the science of climatology" ("A Charter for Climatology," 1953, p. 53, with permission of the World Meteorological Organization).

To put these remarks in their proper context, Thornthwaite was referring to remarks by Professor H. Von Ficker, a dynamic meteorologist and the first president of the IMO Climatology Commission, who maintained that the task of the commission was to create a closer connection between climatology and dynamic meteorology.

The first session of the commission, held in the State Department in Washington, lasted two weeks and was attended by delegates from nineteen member states along with representatives from other United Nations agencies, from the International Union of Geodesy and Geophysics, and from the International Geophysical Union.

In his presidential address, Thornthwaite outlined briefly his views for the future work of the commission: "I believe that the World Meteorological Organization wishes us to consider means of giving climatology a more vital role . . . we need not limit ourselves to the cultivation of climatology for the forecaster's sake. The terms of reference of our commission clearly indicate to me that we are free to encourage the development of climatology independently . . . I feel that we can best promote meteorology by promoting climatology. Therefore, for our commission, I would interpret that directive as meaning 'keep abreast of and promote climatological developments both in the scientific and practical fields'" ("A Charter for Climatology," 1953 pp. 41–42, with permission of the World Meteorological Organization).

In a later section of the same opening address, he provided a general interpretation of his understanding of the terms of reference for CCL: "Our main task will be to see that climatological data are collected and presented in a form so as to be as useful as possible. We must also stress those points of climatology that need research and development. Encouragement of research and the co-ordination of the international aspects of such research are now functions of the WMO and our commission should not hesitate to give advice as to the direction in which climatological research should be carried out. Nor should we close our eyes to the need to promote university training in climatology" ("A Charter for Climatology," 1953, p. 42, with permission of the World Meteorological Organization).

In his closing remarks, Thornthwaite strongly supported the desirability of an international institute for climatology to engage in research and to undertake the training of students at a university in the United States. Without question, he saw the

Laboratory of Climatology as a model for such an institute, and he pointed out the great benefit the staff and students at the laboratory had received from the many foreign visitors who had come to teach and carry out research there. Thornthwaite was ready to have the laboratory serve as this international institute, but nothing developed from this suggestion.

Kenneth Hare, then an associate member of CCL from Canada, summed up the results of the commission's meeting as follows:

The discussions ranged over the whole field of climatological procedure and research. The fifteen resolutions and thirty-eight recommendations that emerged after the final plenary session covered a multitude of subjects. Several working parties were set up to initiate research or inquiry in significant fields. One such group, for example, was established to work with WHO to determine how meteorological data can be applied in the study of climate and health. Another panel was set up to advise bodies such as FAO [Food and Agriculture Organization] on climatic classification. Yet another group was created to study microclimatology. The organization does not yet possess adequate funds to support the work of these working parties, but it is hoped that this obstacle will be overcome very shortly.
The Commission also considered the question of the Arid Zones programs now being carried out by various agencies. The view was expressed that insufficient weight was being given to the climatic factor in many of these programs. The Commission recommended that UNESCO should make climatology the central theme of both the Arid Zone and the proposed Humid Tropics programs for one year, and offered the full cooperation of WMO in these studies. It endorsed the homoclimatic maps of the Arid Zones prepared for UNESCO by Peveril Meigs, and also recommended that member states assist in the Coastal Desert program being initiated by the Arid Zone Commission of IGU [International Geophysical Union]. (Unpublished memo to Thornthwaite from Hare, 1953)

Thornthwaite served a second term as president until January 1957. During the four-year term, the principal emphasis for

CCL as far as Thornthwaite was concerned was the creation of a world climatic atlas. He submitted a draft recommendation for such an atlas for mail vote of commission members in April 1954. In August of that year, in correspondence with G. Swoboda, the acting secretary-general of WMO, he wrote, "I am pleased to learn that the Commission for Climatology has overwhelmingly approved the proposal for a World Climatological Atlas. [The vote was thirty for and five against]. It is clear that four of the five negative votes would have been cast in the positive sense if the word 'guide' had been used instead of 'basis' in Section 3 of the recommendation. If this change is made, the opposition is reduced simply to the unstated grounds for Canada's refusal."

The problem with the use of the word "basis" rather than "guide" was largely one of semantics. If the recommendations were the basis for the preparation of the prospectus, four nations felt that their freedom to modify certain aspects would be severely restricted. The substitution of the word "guide" would allow additional freedom for interpretation of the criteria in particular cases. This change of wording was not a problem for Thornthwaite. The approved recommendation merely called for adopting the principle that a world climatic atlas was desirable, that CCL in consultation with other technical commissions and regional associations of the United Nations be directed to prepare a detailed draft prospective for the atlas, that criteria in the annex to the recommendation be adopted as the basis (guide) for the preparation of the prospectus, and that the atlas be completed on a world basis.

Thornthwaite felt there was widespread interest for such an atlas among other technical commissions and regional associations, and so he proceeded to build support, especially with the Food and Agriculture Organization and UNESCO. He had been made a member of the Committee on Arid Zone Research of UNESCO in 1952. In December 1955, the Working

Group on Climatic Atlases of the WMO met at the laboratory in Centerton, after spending two days at the U.S. Weather Bureau in Suitland, MD. The members of the working group were Thornthwaite (chair), Anders Angström (Sweden), Kenneth Hare (Canada), Stanley P. Jackson (South Africa) and Naftali Rosenan (Israel). The report of that meeting was comprehensive and provided definite guidelines for choosing map parameters, map scales, map projections, units, color tints, printing, paper, binding, page size, and other factors so that maps prepared by different groups would be uniform in style. It was decided that loose-leaf binding would be preferable so that updates of the maps as well as new maps could be added at frequent intervals as more information became available.

Without waiting for the report of the working group, Thornthwaite had started work on some maps of water budget factors in Southwest Asia with the support of UNESCO. He felt that having several maps already available in the style and format of a possible atlas would encourage more support from other agencies and member governments and get the project under way more rapidly. In fact, Thornthwaite received a letter on December 19, 1955 (the day after the end of the working group meeting), from J. Swarbrick, UNESCO Division of Scientific Research, indicating that UNESCO had agreed in principle to publish the water budget maps after further negotiations with Thornthwaite as to how that could best be done. In June 1956, Thornthwaite was appointed United States representative to the Advisory Committee on Arid Zone Research of UNESCO.

In writing to James Swarbrick of UNESCO on January 11, 1956, Thornthwaite showed his enthusiasm about the results achieved by the working group:

> We have produced a climatic atlas project which encompasses national and regional atlases as well. The gist of the project is that

we are following the well-known project for the preparation of the Millionth Map of the world. We have recommended a projection for the land areas of the world which so compromises between conformality and equality of area that all sheets will fit together neatly and yet do little violence to form or area in any part of the map. I believe that there will be about 50 sheets all told in the map on 1:5 million scale.

My proposal to you is that UNESCO undertake to prepare and publish the various maps that will comprise the sheet or sheets that cover Southwest Asia. This suggestion has the virtue that a section of the atlas will be produced for an area of special interest to UNESCO which will not probably be done in the foreseeable future otherwise. These maps could be the first to be published and could thus serve to show what is intended and as models for National Weather Services and others to follow.

Correspondence between Thornthwaite and Swarbrick and others at UNESCO over the next two years dealt mainly with financial arrangements, the preparation of textural materials, the selection of a printer, and the distribution of the map sheets themselves. In October 1957, Thornthwaite attended a meeting of the UNESCO committee in Canberra, Australia, and served as chair of the Australia-UNESCO Symposium on Arid Zone Climatology, and in 1958 he presented a comprehensive paper entitled "Introduction to Arid Zone Climatology." He subsequently lectured in Honolulu, Melbourne, Singapore, Bangkok, Karachi, Tehran, and Geneva. However, it was not until March 20, 1958, that UNESCO drew up a contract authorizing preparation of the water budget maps of Southwest Asia (although they had long been completed), and Thornthwaite signed the agreement on August 18 of that year. On April 13, 1959, he shipped to W. Moller at UNESCO one hundred copies of "Three Water Balance Maps of Southwest Asia," one hundred copies of "Mean Annual Water Surplus" (Red Sea and Persian Gulf sheets), sixty copies of "Mean Annual Potential Evapotranspiration" (Red Sea and Persian Gulf sheets), and sixty copies of "Mean Annual Water Deficit" (Red Sea and Persian Gulf sheets). Forty copies of the last four sheets had already been shipped to Moller. The letter of transmittal ended with the comment, "I shall be interested in learning of any reaction which you might receive from workers in the field concerning the maps and text. We have enjoyed working with UNESCO on this project and hope that sometime it will be possible to extend the compass of this study to make it even more useful for those working with problems of water resources in arid areas."

Although Thornthwaite had hoped to convince WMO through CCL to prepare a significant world climatic atlas, he was successful only in getting support for one small set of water budget maps from another UNESCO technical commission through its Arid Zone Unit. He took the atlas project one step further by preparing similar water budget maps for the eastern United States under a grant from Resources for the Future in Washington, D.C., but the map sheets for Southwest Asia and eastern United States were the only completed. Thornthwaite worked hard to enlist wider support for the mapping project, but his concept was possibly too grand for the member governments at that time.

Thornthwaite stepped down as president of the Commission for Climatology in January 1957, at the end of his second term. He felt satisfied that climatology had taken rapid steps forward in developing independently from meteorology and that significant basic and applied research was well under way in the field. In "The Task Ahead in Climatology" (1957), he concluded,

> In recent years there have been numerous examples of the application of climatological principles in the solution of practical problems in agriculture, industry and defense. I regard this as a wholesome and thoroughly desirable development. However, in order for our science to advance, we must continually add to the store of basic

Thornthwaite presiding at the second meeting of the Commission for Climatology of the World Meteorological Organization in Washington, D.C., January 1957.

climatological knowledge. Applied and fundamental research in climatology must go forward side by side.

It has seemed to me that climatology has gained in stature and risen in prestige during the last four years. In the conclusion of my address four years ago I expressed the hope that climatology might soon be raised from its inferior position in the hierarchy of meteorology. This has come to pass to some degree; climatology has demonstrated its importance and is now far stronger in several meteorological services than formerly. For this progress we may be happy but not complacent. Only by diligent effort can we hold for climatology its newly-won respect. ("The Task Ahead in Climatology," 1957, p. 7, with permission of the World Meteorological Organization)

Warren and Denzil Thornthwaite in Hawaii, en route to Australia in 1957.

Thornthwaite became a well-recognized "citizen of the world" through his travels in connection with WMO, UNESCO, and CCL. In 1957, he participated in a WMO-sponsored regional seminar on the water budget and hydrologic forecasting in Belgrade, Yugoslavia. The seminar brought together climatologists and hydrologists from all countries bordering or near the

Mediterranean Sea. He then went on to a UNESCO meeting on arid zone problems in Karachi, Pakistan. Thornthwaite liked to point out that he had crossed the Atlantic from west to east fifteen times but only 14 times from east to west. Lest one should think this left him in Europe rather than the United States, he would add that on one of those trips he continued on around the world and did not return by way of the Atlantic!

Chapter 9

The Laboratory and the Associates during the 1950s

During the 1950s the Laboratory of Climatology and C. W. Thornthwaite Associates became a mecca for climatologists from all over the world. One of the first visitors in 1950 was Rudolf Geiger, who spent three months at the laboratory lecturing in micrometeorology of the surface layer of the atmosphere. Thornthwaite had realized that a visit to the United States might be beneficial to Geiger at this time, both financially and intellectually. He wrote to invite Geiger to the laboratory, and Geiger answered with the following on March 23, 1950:

Yesterday, I received your kind letter of March 18, and I send you my answer immediately.

I thank you very much indeed for your great kindness. Yes, I am able and willing to come and to give your students a course of lectures in microclimatology, in English, and to advise you in your experimental work as far as I am able to do so. Our summer lectures here are usually finished the last week of July, but I think I can arrange to come at the beginning of July. I would be glad to leave my duties here for about three months. I thank you very much for your kindness in making the necessary arrangements for my travel and accommodation.

As to the language, I was the secretary of the agricultural commission of the International Meteorological Organisation and spoke English frequently during the meetings in Salzburg in 1937. I have no difficulty in being understood by the (English-speaking) Munich Military Government. But I have difficulty in understanding the

USA-English. As soon as I am a fortnight in your country I hope to be out of that awkwardness.

Are the students, who spend their summers in your laboratory, young students, who can be filled with enthusiasm for our beloved science, or will they be older students, already experienced in meteorology?

There is no doubt that the students who were present that summer received, from two world experts, the most comprehensive program in climatology that could be found anywhere in the world. In addition to several students from Johns Hopkins and the Drexel Institute of Technology, there were several foreign students and three U.S. Air Force officers in residence at the laboratory. Other foreign lecturers invited to Centerton during 1950 included Maung Hla, assistant director of the Meteorological Service of Burma, and C. H. B. Priestley, officer in charge of the Section of Meteorological Physics of the Commonwealth Scientific and Industrial Research Organization, Australia.

Ben Garnier, a New Zealand climatologist, was another visitor to the lab in 1951. In 1946 he had published a paper entitled "The Climates of New Zealand According to Thornthwaite's Classification." George Jobbeous, the father of New Zealand geography, sent a draft of this paper to John Leighly; he then showed it to Thornthwaite, who wrote a letter to Garnier containing some helpful criticism and suggestions and inviting him to visit the laboratory. Ben Garnier, now retired from the Geography Department of McGill University in Montreal, recalls his first meeting with Thornthwaite in a letter to the authors:

> I well remember arriving in Centerton in June, 1951 after 14 hours in a stratocruiser London/New York via Iceland, four hours in customs, a train ride to Philadelphia and bus to Centerton, his kind, smiling, friendly welcome. He settled me in Carl Hohn's house where I stayed and then said "You must be hungry after such a long journey"

Rudolf Geiger, famous German microclimatologist, lecturing to staff and students of the Laboratory of Climatology, Seabrook, N.J., summer 1950. (Photography by E. Taubert, reproduced with kind permission of the Seabrook Farms Co.)

and took me off to the English pub [the Centerton Inn] and ordered a steak larger than any I've seen before or since! I managed about half of it; but the rest wasn't wasted, but (to my British astonishment) went home with Warren in a brown paper bag for his dog! I guess that little episode set for me the love of Warren, a kind and warm person, despite a firm and uncompromising professional/scientific mind and judgement. He invited me to his house several times and I enjoyed his company, that of his wife and I think I met one daughter and of course the dog! It was a lovely family house and he made me feel so welcome.

One day early in my visit he drove me to see various things (at least with that intention), and we saw a black man with knife cuts bleeding by the road. Warren stopped at once and picked him up,

Thornthwaite explains modifications incorporated into the standard three-cup anemometer used in an earlier microclimatic study to German climatologists Rudolf Geiger and Heinz Lettau at the Laboratory of Climatology in summer 1950. (Photography by E. Taubert, reproduced with kind permission of the Seabrook Farms Co.)

and off to the hospital. The man said Warren's was the first car to stop: several others had passed and he had been there nearly an hour. Warren not only took the man to the hospital but stayed until he was sure he was going to get proper treatment.

Warren went out of his way to help all. He saw to it that I visited Johns Hopkins and met in particular Douglas H. K. Lee which opened for me an interest in physiological climatology. He also contacted John Leighly personally to meet me when I went off for some time in San Francisco en route to a visit to Sydney. Also he was full of good sense and practical advice to me, the novice in evapotranspiration.

During the early 1950s at the laboratory, the research being conducted for the U.S. Air Force on micrometeorology of the surface layer of the atmosphere became more sophisticated and comprehensive. Wind profile measurements became more exact, resulting in improved computations of the climatic fluxes near the ground. The theory of turbulent motion in the lowest layers of the atmosphere was being rapidly refined. The instrument section of the laboratory produced new micrometeorological sensors able to measure small, short-period fluctuations in wind, moisture, and temperature at various heights above the ground. It was generally recognized that the micrometeorological studies being carried out at the laboratory under Thornthwaite's guidance were at the leading edge of the field and that the instruments being developed were better than any that had been available previously. In 1951, work also began on a contract with the Office of Naval Research (ONR) to prepare detailed, large-scale climatic maps of Africa.

Consulting work for Seabrook Farms began to decline as the various activities of the laboratory increased. Thornthwaite saw commercial possibilities in his Cropmeter, the slide rule device that converted the civil calendar to a climatic calendar and allowed one to schedule the planting and harvesting of a wide range of garden vegetables. Much time was spent not only on trying to devise the most usable version of the Cropmeter but also on gathering the crop development data needed to accompany it.

Thornthwaite continued to speak about agricultural climatology at Seabrook Farms to various agricultural groups, and his comments were quoted and discussed in many farm-oriented publications. In 1952 he contributed an entry on drought to the Encyclopedia Britannica, published a paper entitled "Grassland Climates," and prepared another entitled "Evapotranspiration in the Hydrologic Cycle," which was printed in a report of the House of Representatives Interior and Insular Affairs Committee.

Thornthwaite was honored in 1952 when the Association of American Geographers (AAG) gave him its Outstanding Achievement Award. (This form of honors award preceded the current award now given annually by the association to the most worthy geographers in the world. In those early years, the association gave two sets of honors awards, one for meritorious achievement, and one for outstanding achievement. In 1976, the association decided to combine these two awards into one, identified as AAG Honors, to dispel the impression that one class of honors was superior to the other).

As Thornthwaite's reputation grew during the early part of the decade, he was asked to serve on the Geography Advisory Committee to the Office of Naval Research of the National Research Council, the Subcommittee on Agroclimatology of the Committee on Plant and Crop Ecology of the National Research Council, the Advisory Panel for Earth Sciences of the National Science Foundation, and the Advisory Committee on Vegetation and Water Yield of the Conservation Foundation. He was also a member of the New Jersey Water Pollution Committee of the Tri-State Packers' Association, Inc., and served as a consultant to the United States Army Signal Corps, Seabrook Farms Co., Phillips Canning Co., the H. J. Heinz Co., and the Refrigeration Research Foundation. These were busy and exciting years.

Foreign visitors to the laboratory increased each year. T. Sekiguti of the Central Meteorological Observatory in Tokyo came for a nine-month stay in 1952 and completed a very fine series of maps of Japan, Korea, and Formosa (Taiwan) on a scale of 1:1,000,000. Less detailed map series of Australia, China, India, and the Middle East were prepared by other visitors: J. Gentilli of the University of Western Australia; H. von Wissmann, dean of the Department of Geography of the Univer-

Warren Thornthwaite, A. Austin Miller (England), and Takeshi Sekiguti (Japan) discuss problems of climate mapping at the Laboratory of Climatology in 1952. (Photography by E. Taubert, reproduced with kind permission of the Seabrook Farms Co.)

sity of Tübingen; V. P. Subrahmanyam of the Andhra University, Waltair, India; and E. Naftali Rosenan, assistant director of the Israel Meteorological Service. In addition to their work on the climatic maps, the visiting scientists gave lectures and seminars at the laboratory. Other famous scientists who spent some time at the laboratory during 1952 included Anders Angström from Sweden; C. E. P. Brooks, S. W. Wooldridge, and A. Austin Miller from England; H. Fukuda from Japan; A. Contreras Arias from Mexico; A. H. Kamph from Denmark; F. Albani from Argentina; K. Troll and H. Flohn from Germany; L. J. L. Deij and H. ten Kate from the Netherlands; and A. A. Wilcock from Australia.

The U.S. Air Force's Great Plains Turbulence Field

Expedition to O'Neill, Nebr., was begun in 1953. Along with research groups from a number of government agencies and other universities, the laboratory sent a fully equipped micrometeorological trailer, masts, profile measuring sensors, and a team of seven scientists and technicians to the flat, open site in central Nebraska for the period July 24 to September 11, 1953, to make detailed observations of all factors of the turbulent structure of the lower atmosphere using as many different techniques and sensors as possible. The combined salary for the seven laboratory staff for the month and a half period came to $2,552, and each individual received a living allowance of $8 per day. Clearly, this was a spartan operation being carried out by dedicated personnel.

Thornthwaite continued his prodigious output of publications. In 1952 his presidential address for the Commission for Climatology "A Charter for Climatology," was published in the *WMO Bulletin;* an article entitled "Operations Research in Agriculture" appeared in the *Journal of the Operations Research Society of America;* "Klima-Kalender und Pflanzenwachstum" was published in *Die Umschau;* and "The Water Balance in Arid and Semiarid Climates" was included in *Desert Research,* the proceedings of an international symposium held in Jerusalem. During 1953, foreign visitors included C. C. Wallén from Sweden; Q. Hamid from Pakistan; M. X. Thaller from Israel; Farag Mohamed Ali from Egypt; Howard Penman from Rothamsted, England; and Gordon Manley from the University of London. Woodrow Jacobs and Helmut Landsberg, two outstanding American climatologists, also visited the laboratory.

During 1954 an agreement was worked out with the Drexel Institute of Technology, and the laboratory was affiliated with Drexel for the next five years. A second change involved the laboratory's move to Centerton, N.J., where it was to remain for the rest of Thornthwaite's life. A third change was the termination of the long-term contract with the U.S. Air Force on the study of micrometeorology of the surface layer of the atmosphere. A new, three-year study with the Air Force on the time-height variations of micrometeorological factors during radiation fog was begun, as was a study with the Office of Naval Research on the heat and water balance of the earth. This was the first of many studies with ONR that provided many years of fruitful research collaboration. Also in 1954, Maurice Halstead, who had been the assistant director of the laboratory since its beginning in 1948, left in the summer to accept a research position in meteorology at Texas A&M University.

"Topoclimatology"—possibly Thornthwaite's most significant publication since his 1948 climate classification paper—appeared in 1954 as well. It had been presented to a Royal Meteorological Society gathering in Toronto in late 1953 and was published in the proceedings of the conference. Although it was not widely circulated in published form, it had a considerable impact on climatology because it clearly expressed Thornthwaite's view that the study of climatology had to include the understanding and mapping of factors of the earth's surface that affect the heat and water balance—namely topography, albedo, slope, aspect, vegetation, and surface roughness.

Anders Ångström of Sweden was a frequent visitor to the lab, and he arrived for a four-month visit in May 1955. He contributed to the work on radiation measurements and the development of improved radiometers, and he also gave lectures to the staff and students at the laboratory. Thornthwaite was an invited participant in a 1955 symposium in Princeton celebrating George Perkins Marsh's classic book, *Man and Nature*. Thornthwaite's contribution was a paper entitled "Modification of Rural Microclimates," which was published the following year in *Man's Role in Changing the Face of the Earth*, the symposium proceedings. Other publications during this period included "Climatic Classification in Forestry (1955, with

F. K. Hare) in *Unasylva,* "Climatology and Irrigation Scheduling" (1955, with J. R. Mather) in *Weekly Weather and Crop Bulletin,* and "The Water Balance" (1955, with J. R. Mather) in *Publications in Climatology.* This third publication summarized the work that had been done on the water balance since it was first reported in *Geographical Review* in 1948 and discussed some important modifications of the bookkeeping procedure.

Although Thornthwaite generally resisted small modifications to the formulation of the water balance procedure, he accepted the changes summarized in "The Water Balance" because they seemed to make the procedure more rational. His standard answer to those who suggested specific modifications of the procedure was expressed in a letter to Canadian climatologist and hydrologist Arleigh Laycock in June 1955: "We certainly realize that the present evapotranspiration formula is inadequate under certain special situations and that minor regional modifications would make it more applicable locally. However, we are adverse [sic] to making these types of changes since they would lead to a larger number of evapotranspiration expressions, all of limited usefulness. Rather, we are trying to obtain a better picture of the physics of evapotranspiration and to limit any modifications of the formula itself to those which are physically sound and universally applicable."

Under contract with the U.S. Air Force, the laboratory carried out research during 1955–1956 on microclimatic conditions at the time of occurrence of radiation fog, on the estimation of soil tractionability from climatic data, and on the preparation of maps of soil moisture and the state of the ground for selected areas. Under contract from the Office of Naval Research, research was carried out on the moisture balance of the earth and on the inflow and outflow of water in the Mediterranean Basin. In mid-August of 1955, Thornthwaite visited Point Barrow, Alaska, with members of the Office of Naval Research to advise on climatic problems and to initiate the development of a thorough study of the microclimate of that Arctic region, on which the laboratory began work during 1956.

Each year brought more scientists from all parts of the world to the laboratory to discuss problems of mutual concern. Some remained for several weeks or months and gave lectures and seminars on subjects of geographic and climatic importance. In 1955 these included D. A. Davies and Oliver M. Ashford of the World Meteorological Organization, Geneva, Switzerland; V. H. Guerrini of the Irish Meteorological Service; Stanley P. Jackson of Witwatersrand University, Johannesburg, South Africa; Naftali Rosenan, assistant director of the Israel Meteorological Service; Hermann von Wissmann, University of Tübingen, Germany; H. O. Slatyer, Commonwealth Scientific and Industrial Research Organization, Canberra, Australia; E. Kraus, Snowy Mountain Hydroelectric Association, Cooma, Australia; Barket-Ullah Khan of Tando Jam, Pakistan; and Augustine Y. M. Yao of the Chinese Meteorological Service, Formosa.

During 1956 the research project dealing with microclimatic conditions at the time of radiation fog, which had been undertaken for the U.S. Air Force in 1954, was completed. Two research studies supported by the U.S. Navy, dealing with the water balance of the earth and of the Mediterranean Basin, were also completed. The work on soil tractionability supported by the U.S. Air Force was continued, and a new study on the water balance of the Delaware Basin was begun under sponsorship of the Office of Naval Research.

The laboratory's program to produce maps of climatic water budget factors of different areas of the world produced thirty-one large-scale maps for publication. At the same time, a study was undertaken to determine the change in the monthly amount of water stored in the soil in all land areas of the globe. This study, which resulted in values of storage based on a one degree latitude by one degree longitude grid, used data from fifteen thousand stations worldwide and provided a realistic check on a study of changes in sea level done by the Scripps Institute in La Jolla, Calif.

In 1956, foreign visitors to the laboratory included John Phillips, University College, Gold Coast, Africa; Herman Flohn, Frankfurt, Germany; Walter Nebiker, Svenn Orvig, and Ian Jackson, McGill University, Montreal, Canada; Halis Alagoz, Ankara, Turkey; Leslie Curry, Auckland University, New Zealand; C. C. Wallén, Swedish Meteorological and Hydrologic Service; Oliver Ashford, WMO, Geneva; James Swarbrick, UNESCO, Paris; Karl Keil, Bad Kissingen, Germany; Contreras Arias, director of Geography and Meteorology for the Mexican government; and C. L. Godske, University of Bergen, Norway.

The research study for the Office of Naval Research on the water balance of the Delaware Basin was continued during 1957, though the nature of the research program shifted from a study of the average climatic water balance conditions over the basin and the small subbasins within the Delaware River basin to a detailed micrometeorologic study of the flux of heat and moisture at the earth's surface. Such a study was begun at the laboratory's experimental field in Centerton with the installation of a complete micrometeorological observing station. In addition to measurements of solar and net radiation, observations of the profiles of air temperature, soil temperature, soil heat flux, dew point temperature, and wind velocity near the surface were also made, and the precipitation and evapotranspiration were measured.

In May 1958, a new research contract with the U.S. Air Force Office of Scientific Research was begun. This research project involved a study of the redistribution of radioactive strontium-90 in the soil after it had fallen from the atmosphere. Because of the importance and complexity of the problem, it was expected that several years would be necessary to complete this research commitment. Research for private corporations also increased and included studies of hazards associated with the operation of nuclear reactors, agricultural studies for a large, mid-South cotton plantation, and studies of wastewater disposal for industrial plants in the United States and Canada.

In September 1958, work started on a new research study for the U.S. Signal Research and Development Laboratory on the measurement of vertical winds over typical terrain. The study had two phases: the first involved the development of an instrument to measure the updrafts and downdrafts associated with turbulent eddies in the air, and the second involved the use of the instrument to make measurements at heights up to five hundred feet over different types of land surfaces.

The research studies of the heat and moisture balance of the Arctic conducted at Point Barrow, Alaska, were completed in October 1958. The study had lasted two and one-half years and included eighteen months of actual fieldwork at the Arctic Research Laboratory in Point Barrow. A comprehensive report on the study was published at the end of the contract work. The research study of the water balance of Southwest Asia, which had been undertaken for UNESCO, was also completed in 1958, and a report was issued that included six large-scale maps of the distribution of potential evapotranspiration, water surplus, and water deficit.

Research for private industries involved considerable time during 1958. These studies included site evaluations prior to the building of nuclear reactors in Puerto Rico and Tehran, Iran, as well as a resurvey of the nuclear research reactor site near Princeton, N.J., before full-scale operation of the reactor began; agricultural climatology studies for a large, mid-South cotton plantation; and the design and construction of wastewater disposal systems for industries in several places in the United States.

Warren Thornthwaite receives the Cullum Medal of the American Geographical Society from President Walter Wood in New York City in 1959.

Thornthwaite received a major honor in 1959, when he was awarded the prestigious Cullum Geographical Medal of the American Geographical Society. The Cullum medal is given to those who distinguish themselves by geographical discoveries or the advancement of geographical science. Previous medal recipients included such well-known explorers as Robert E. Peary, Fridtjof Nansen, Robert F. Scott, and Sir Ernest Henry Shackleton. The citation read at the American Geographical Society meeting at which the Cullum medal was presented was as follows:

The career of Charles Warren Thornthwaite—climatologist, ecologist, geographer—has displayed throughout two marked characteristics: originality of thought and effective application of theory.
After eleven productive years as chief of the Climatic and Physiographic Division of the Soil Conservation Service—years that co-

incided in part with the disastrous drought that produced the Dust Bowl—Dr. Thornthwaite in 1948 established, and has since directed, the Laboratory of Climatology in New Jersey, first under the aegis of The Johns Hopkins University and later under that of the Drexel Institute of Technology. This center has in the intervening decade become a lodestone that draws students and researchers from all parts of the world.

From his years with the Laboratory of Climatology has come the fine work for which Dr. Thornthwaite is best known. Here he formulated his "Rational System for the Classification of Climates," first published in the *Geographical Review* and thereafter adopted and applied by scientists in many different countries. Here he developed his important concepts relating to water balance and evapotranspiration. Here, too, he initiated experimental studies of the effects of microclimatic conditions on plant growth and crop yields. These studies proved highly successful in their application to commercial market gardening.

The realm of science is the richer for the ideas, experiments, and writings of Charles Warren Thornthwaite, and it is with justifiable pride that the American Geographical Society presents to him its Cullum Geographical Medal in recognition of the creative contributions he has made to the world geographers live and work in. (Reprinted with permission of the American Geographical Society)

Walter Wood, president: William A. Rockefeller and J. Clawson Roop, vice presidents; and Charles B. Hitchcock, director of the American Geographical Society, all attended the presentation. Lloyd V. Berkner gave an after-dinner talk entitled "Geography and Space Exploration." The dinner menu (including Tournedos de Boeuf Forestière and for dessert Couronne de Glace Pêches Flambées), preserved in the Thornthwaite files, indicates that it was an elegant occasion and a fitting conclusion to the most exciting decade of Thornthwaite's life.

Chapter 10

The Last Years

For Thornthwaite, the 1960s began with work on the development of a variety of sensitive microclimatic instruments, with studies for the Office of Naval Research, and with consulting on different wastewater disposal projects. He continued to travel widely and to receive honors for his contributions to climatology. He was chosen honorary president of the Association of American Geographers for 1960–1961, and in August 1961 he presented his presidential address at the association's fifty-seventh annual meeting in East Lansing, Michigan. Using "The Task Ahead" as a title, Thornthwaite launched a fairly strong attack on the regional geography of the day:

> It appears that the great majority of those who call themselves geographers are professors and teachers. They are therefore mainly concerned with the presentation of material to students. For the most part the title "geographer" means University professor, the center of activity of the geographer is the classroom, and a "regional specialist" is one who can organize and teach a regional course in a creditable fashion in a college classroom. The research that would thus be expected of the geographer consists of seeking out and assembling the course material from a variety of sources, organizing it according to some logical pattern, designing accompanying maps, and selecting relevant photographic illustrations. The skills involved are the ones that would enable him to carry out this program. They would include ability to write in acceptable English, read foreign language sources, and have some knowledge of maps.
> That growing numbers of geographers entertain different views as

to the content and objectives of geography and as to the training required for productive work in the field is attested by the increasing emphasis on quantitative methods. This new development has aroused a certain uneasiness among the regionalists and may have put them on the defensive. It is admitted that "the proponents of the modern methods of quantitative analysis have raised geographical study to a higher level of sophistication," but "the focus of geographical attention [remains] on the characteristics of particular places." Almost sorrowfully, we are asked to consider the question: "Where is the current trend toward increasing quantification taking us?"

Thus, arising at the side of Leighly, who charged that the physical problems of the earth are a part of the responsibility of geography, are the geographers who see that economic, social, and cultural problems can have a depth not to be reached by the so-called "standard" geographical techniques. I am sure that Leighly did not exclude those problems when he said, "The land, the sky, and the water confront us with questions wherever we look at them." The difficulty of course is that to get to the bottom of any basic question one needs a different training than is offered for prospective geography teachers by present-day Departments.

There has never been any doubt that to do acceptable work in any of the physical branches of geography such as geomorphology, climatology, hydrology, or pedology, mastery of mathematics and physics, together with an extensive knowledge of chemistry, are required. The present work of the regional scientists being published in the *Journal of Regional Science*, some of which is outstanding, brings into focus the need for genuine competence in mathematics in regional geography as well. I have been urging this need on the part of geographers for many years. One of my colleagues of a generation ago dismissed the matter by saying, "When I have a mathematical problem I will hire a mathematician to solve it for me." Needless to say, he never had a mathematical problem because he could not know what mathematics could do for him. During recent years a few young geographers have demonstrated the power of modern mathematical and physical science by using them in the solution of problems in economic and cultural geography and in various population problems. I must admit that some nonsense has appeared also. We have much evidence, however, that the student who has mastered any of the systematic branches of geography, and who has a solid foundation in physical science to back it up, makes a better geographer. These are the ones who can perform creditably when they are required to do something other than train the teachers of undergraduate courses in geography. ("The Task Ahead," *Annals of the Association of American Geographers,* 51 [1961]: 347–48, with permission of the Association of American Geographers)

Although the presidential term in the Association of American Geographers is only one year, and thus there is little opportunity for an individual to influence the course of the discipline, many geographers feel that Thornthwaite's ringing appeal in his presidential address for a more quantitative and rigorous approach to the training of geographers had a significant influence on the training of physical geographers in the decades following its publication. This belief was stated by Melvin Marcus in his presidential address to the association in 1979: "Warren Thornthwaite reinforced our belief that 'The Task Ahead' required rigorous attention to process and understanding of environmental systems. Considering that most persons attending the East Lansing banquet (mostly human geographers) claimed to have been bored and/or unable to understand his address, it is worth noting that eighteen subsequent years of American physical geography have followed the essential spirit and formula of Thornthwaite's mandate. (*Annals of the Association of American Geographers* 69 [1979]: 525, with permission of the Association of American Geographers)

Thornthwaite had other problems that were of more immediate concern. These were problems of health, both his own and those of his wife, Denzil. Thornthwaite did not allow these problems to interfere with trips and work activities unless absolutely necessary, and he never wanted to admit to any illness. For example, in the mid-1950s, Thornthwaite came down with mumps, a fairly serious illness for one approaching sixty. Although those at the laboratory who had never had mumps were not happy to see him at work, he refused to stay home and rest and worked nearly every day.

Denzil had had a malignant tumor removed from her lung in 1954, and in the spring of 1960, she returned to the hospital to have the lower lobe of her right lung removed because of a recurrence of the malignant tumor. Her recovery was very slow this time, and because she seemed to feel that the end was approaching, she did not want her daughters or husband to leave on trips. However, Thornthwaite and the three girls left to go to Germany and Sweden to attend the Nineteenth International Geographical Congress in 1960. Denzil was left in the care of Doug and Murphy Carter (Doug was a colleague at the laboratory), and in Thornthwaite's own words her improvement was "positively miraculous." Thornthwaite noted that in early October "she drove by herself to [the nearby town of] Elmer for the first time since last April" (letter of October 10, 1960, from Thornthwaite to F. K. Hare).

During 1961, Thornthwaite experienced back pains, but he had had these for much of his life. For some time he had mentioned the difficulty he had in walking on pavement for any distance though he did not have this problem on unpaved surfaces. Denzil had increasing problems with breathing and was not as active as she had been in her younger years, but the two did manage several trips together. However, Denzil's health deteriorated rapidly in the spring of 1962, and she died on July 29, from carcinomatosis, at age sixty-three. This was a tremendous blow to Thornthwaite. He suffered increasing back pains during the last few months of Denzil's life without allowing it to interfere with his work schedule, but once she had gone, his will to fight the increasing pain apparently ebbed.

To honor Denzil, Thornthwaite created a scholarship fund in her name at Central Michigan University, the undergraduate college they both had attended. He visited the university on

The Thornthwaite family together for Christmas 1961. Standing, left to right, are daughters Dorothy, Elizabeth, and Sally.

Labor Day, 1962, to make arrangements for the scholarship fund. His letter to President J. W. Foust of Central Michigan University on October 19, 1962, described his health problems:

Dear President Foust:

When I arrived home from my visit to Mt. Pleasant last Labor

Day my back had become so bad that I found myself completely out of action. A few days later I was brought to the hospital in an ambulance. An operation has been performed—a collapsed disk was removed, and I have been on my back in bed for an eternity. There was a setback, a blood clot in the large vein of the left leg, but finally I am up and able to begin taking care of my private correspondence. The doctor makes no promises as to when I will be discharged. However, I no longer have any pain and do not mind hospital life. . . . I am scheduled to be at the University of Chicago the last week of November. If I am able to fulfill my obligation there and have no trouble, I would like to stop off again in Mount Pleasant on my return. I would like to say "hello" to you but would have no reason to ask for any of your time.

Although he had recovered enough to make a business trip to California in early December, he was unable to attend the presentation of the check to Central Michigan University that marked the initiation of the scholarship. His eldest daughter, Elizabeth, went to Mount Pleasant, Michigan, to make the presentation in his place. Following Warren's death in 1963, at the request of his daughters, this scholarship fund was renamed the Charles Warren and Denzil Slentz Thornthwaite Memorial Scholarship.

Recognizing Warren Thornthwaite's unique contributions to climatology and geography worldwide, Central Michigan University sought to honor possibly its most outstanding graduate and chose to award him an honorary doctorate in June 1963. Unfortunately, Thornthwaite had to receive this award in absentia; the ceremony took place just nine days before his death.

The citation accompanying the awarding of the degree, the last honor Thornthwaite was to receive during his lifetime, is reprinted below.

Dr. Charles Warren Thornthwaite, Bachelor of Arts, Central State Teachers College, 1922, and Doctor of Philosophy, University of California, 1930, has a long and distinguished list of contributions to the science of climatology and a no less distinguished array of honors bestowed upon him by his fellow scientists.

Universities which have sought his scholarly research and instruction by appointments as full professor include the University of Maryland, Johns Hopkins University, which asked him to establish and direct its Laboratory of Climatology, Drexel Institute of Technology, and the University of Chicago. He has also served as Chief, Climatic and Physiographic Division, United States Soil Conservation Service, as lecturer at numerous international conferences, and as consultant to fourteen foreign nations in the Americas, Europe, and Asia.

He is an Honorary Fellow in the American Geographical Society, a contributor to its *Geographical Review,* and the recipient in 1959 of its much coveted Cullum Medal.

He is a Fellow of the American Geophysical Union, having served as Vice President of its Meteorological Section and then as its President.

For seven years he was President of the Commission for Climatology of the World Meteorological Organization, being elected to that office in absentia to the surprise of no one but himself.

From 1955 to 1958 as the United States Member of UNESCO's Committee on Arid Zone Research, he became immersed in the cauldron of the Middle East, where he, a physical geographer, found himself as much interested in the human problems of the people as in their climate.

In 1956 he was Chairman of the Australia-UNESCO Symposium on Arid Zone Research in Canberra, Australia, and in 1960 the official Delegate of the National Academy of Science, National Research Council, to the Tenth General Assembly and Nineteenth Congress of the International Geographical Union in Stockholm, Sweden.

In its feature section, "Personality Portrait," the *Saturday Review* for February 7, 1959, credits Dr. Thornthwaite with being the foremost authority on the micromechanics of the water balance and a pioneer in the science of Microclimatology.

Whether the name of the man we honor, Thornthwaite, which in the early days of our language meant a dweller by the thornbushes near a forest clearing, in any way symbolizes his life's work, his science has helped to transform the dust bowls of the Western States into verdant plains and has given hope to the nomads of the deserts of the Middle East.

Central Michigan University is indeed proud to confer upon Charles Warren Thornthwaite the honorary degree, Doctor of Science.

June 2, 1963, Judson W. Foust, President
Mount Pleasant, Michigan, Central Michigan University

The back pains returned and had become severe by the end of December 1962. He entered the hospital again for observations early in January 1963, but he soon returned home and was confined to complete bed rest. His condition did not respond to this treatment, and he returned to the hospital on February 23 and had a hemilaminectomy performed on February 24. A biopsy taken at that time apparently gave no evidence of a malignant growth. He appeared to be recovering from the operation, but it soon became obvious that all was not well because he was unable to sit up. On March 20 he was placed in a body cast and remained in it until May 7. His condition did not improve, and he continued to lose weight. About May 14, Thornthwaite informed his associates that the doctor had told him he had bone cancer and that he had three to six months to live. He died on June 11, 1963. Both he and his wife had died of cancer less than a year apart.

Chapter 11

Thornthwaite's Legacy

Warren Thornthwaite was born in 1899 and died, while still in his scientific prime, at age sixty-four. Longtime colleague and friend Ken Hare wrote the following obituary in the *Geographical Review:*

Warren Thornthwaite was not an easy man. He did not suffer fools gladly, nor did he take kindly to criticism. Much of his scientific writing does scant justice to the genius that inspired it. He never submitted to the rigor and discipline required by a high scientific style. Yet the brilliance and originality of his insight were unquestionable to those who knew him best. There were many who resented his scalding castigations of his critics. But there were many more who loved him, and they will long mourn his passing. Foremost among these will be the army of professional brethren who were royally entertained at Centerton. Prominent, too, will be the many Japanese and European climatologists who, after World War II, were helped back on their feet by his remarkable generosity. Without doubt he was the most influential climatologist of his generation. (Hare, *Geographical Review,* 1963, with permission of the American Geographical Society, 53:597)

Another obituary, in the *Annals of the Association of American Geographers,* was written by John Leighly. Leighly's three-page obituary indicates that he did not judge very accurately Thornthwaite's contribution to science, nor the popularity of his water balance writings. Leighly wrote, "To judge from Thornthwaite's repeated exposition of his procedure for estimating the water balance of terrestrial surfaces from observed precipitation and temperature, he looked upon this as his most important achievement. Yet it was seriously questioned in his lifetime and may now be considered obsolete" (*Annals of the Association of American Geographers* [54] 1964:615, with permission of the Association of American Geographers).

After Thornthwaite's death, his colleagues continued to discuss his scientific achievements. A symposium in his honor was held during the International Geographical Congress in Montreal in 1972, with papers by many of Thornthwaite's friends and associates. L. A. Ramdas, scientist emeritus at the National Physical Laboratory at New Delhi, India, was invited to the Congress. He wrote,

> I would like to take this opportunity to record some of my reminiscences of Dr. Thornthwaite. For nearly quarter of a century from 1933 onward he was like an intimate "pen-friend" with whom I exchanged ideas and publications on many topics of mutual interest. It was only in 1956 that we met each other at Canberra in Australia where we were both attending the UNESCO Conference on Climatology and Micro-Climatology. Dr. Thornthwaite was presiding over this Conference. After the conference many of the leading climatologists were invited by the CSIRO [Commonwealth Scientific and Industrial Research Organization] at Melbourne to join in a high-level symposium to discuss scientific problems and latest advances in Climatology. So, Dr. Rudolf Geiger, myself and Dr. Dzerdzeevski of USSR were also there. The entire party of visiting scientists was invited to a special meeting of the Australian Meteorological Society one evening at Melbourne. Dr. Thornthwaite was requested to take the chair, by the President of the above Society. In his opening remarks Dr. Thornthwaite said that he was happy to state that three of the leading climatologists of the world were present and that he would introduce them one by one and ask them to stand up and bow to the audience. He introduced Dr. Rudolf Geiger and Dr. Dzerdzeevski and then myself. In introducing me he mentioned that he and I were known to each other only by correspondence over many years and he said: "I was expecting to meet a formidable looking very

elderly person with an imposing beard. Look at him, he is a mere boy!" I bowed to the audience feeling ever so shy for my alleged youthful appearance!

We next met at one of the annual sessions of the UNESCO Arid Zone Advisory Committee, most probably in Paris. While I was visiting many Universities in USA in the fall of 1960 and was lecturing at Wisconsin, Dr. Thornthwaite attended an evening talk by me and cordially invited me to visit Centerton and I promised to do so. Later, in November of 1960 he was good enough to meet me at the airport in Philadelphia and drive me to Centerton. After lunch at his daughter's place we went over to Dr. Thornthwaite's home where we spent a very interesting time seeing our many slides illustrating many scientific results in Climatology and Micro-climatology. After spending the night there he took me next morning to the Office of Thornthwaite Associates where I met all of you and later attended the lunch. You may recall that after the lunch I was called on to address the entire staff on my tour in USA as Visiting Foreign Scientist and all that I had seen at the various institutions already visited by me. I left the same evening for New York.

We met again and for the last time, as it happened, at the Ten Year Planning Conference on Atmospheric Physics held in the summer of 1961 under the Chairmanship of Dr. S. Petterssen at Boston at the headquarters of the American Meteorological Society.

Rudolf Geiger, retired and living in Munich, was, of course, invited to the symposium. His answer to one of the authors is reproduced, below.

I thank you very much for your kind letter concerning the Congress at Montreal. You know that I am not able to come to Montreal myself. I am sorry indeed that I cannot participate in the two sessions in honor of my most venerated C. W. Thornthwaite. I have marked the date in my calendar and all day and during the luncheon my thoughts will be with the participants at Montreal.

The amiable character of C. Warren Thornthwaite was the great impression of my post-war years. I saw him the first time in my life as I arrived July 6, 1950, at New York. Among the 1,600 passengers of the *Queen Mary* I was dismissed as the 1597th because I was a German. As I looked for my luggage under the sign-board "G"—

there he stood, Dr. Thornthwaite, asking me: "Are you Professor Geiger?" Three hours and 45 minutes he had been waiting for me at the customs office. It was my first opportunity to experience his patience and his unusual kindness.

He had informed himself about our sufferings during the wartime years. As we stood in the Arlington National Cemetery at the Tomb of Unknown Soldier, he laid his arm round my shoulder, without saying a word, remembering the son I lost in the war.

Certainly, in the morning and afternoon session you will be discussing the manifold scientific investigations of Dr. C. W. Thornthwaite: on the water balance of the earth and the results and new prospects he opened everywhere in that field of climatology with practical applications. But what is the worth of the richest scientific work in comparison with the human lovable character C. W. Thornthwaite was endowed with?

If anyone at the meeting will remember me, please give them my kindest regards.

Another famous climatologist who was invited to the Thornthwaite Memorial Session was M. I. Budyko of the Main Geophysical Institute of St. Petersburg, Russia. Dr. Budyko replied,

I am very sorry that I cannot accept your invitation and take part in the luncheon planned to honour the work of C. W. Thornthwaite, since I shall not participate in the International Geographical Congress.

I would like to ask you to tell the participants in the luncheon that, in my opinion, the works by Prof. Thornthwaite were of great importance for creating the present climatology and opened new perspectives for developing this science.

My personal acquaintance with Prof. Thornthwaite left with me unforgettable recollections of him as a personality and a scientist.

Thornthwaite had created two organizations when he moved from government service to Seabrook. One was the Laboratory of Climatology, which he hoped would fulfill his long-cherished dream of an institute of climatology attached to a first-rate university where significant research could be initiated and graduate students trained in the science of climatology. The second was C. W. Thornthwaite Associates which was, at first, merely Thornthwaite himself acting as a consultant. Later, it became a partnership with his brother-in-law, Floyd Slentz, and finally a not-for-profit corporation that handled federal research contracts obtained by the laboratory.

For the first six years of its existence, the laboratory was the Johns Hopkins University Laboratory of Climatology; then, following the separation from Johns Hopkins in 1954, it became the Drexel Institute of Technology Laboratory of Climatology. This association lasted five years and was brought to an end by Thornthwaite himself when it became clear that government contracts could be written with C. W. Thornthwaite Associates. In this way, the overhead on the contracts would help to defray some expenses incurred by C. W. Thornthwaite Associates, which owned the laboratory buildings. Thus, the last name change was to C. W. Thornthwaite Associates Laboratory of Climatology.

When C. W. Thornthwaite Associates was organized as a not-for-profit corporation, stock representing ownership of the corporation was issued to the five officers: Thornthwaite, president; John R. Mather, principal research scientist; June Yoshioka, secretary/treasurer; Bill Superior, head of the instrument development laboratory; and Don Parmelee, head of the wastewater disposal program. Upon his death, Thornthwaite's stock was bought back by Thornthwaite Associates as stipulated in the original papers of incorporation. Business continued but hardly as usual. The laboratory had several government research contracts to complete, the demand for the sensitive micrometeorological instruments was brisk, and many wastewater disposal jobs remained to be undertaken.

There was little doubt in the officers' minds that the work of the associates could continue, considering that its reputation, established by Thornthwaite during his lifetime, was significant. However, the officers had their individual careers and interests to consider. In 1972, Superior bought the organization from the remaining stockholders and continued the instrumentation work of the associates. Mather moved to the University of Delaware to institute the Department of Geography, Parmelee took the wastewater disposal work from the associates and continued as a private consultant, and June Yoshioka took a position as an officer in a local utility organization.

Thornthwaite had started a publications series in 1948 during the first year of the Laboratory's operation. Called *Publications in Climatology,* it served as an outlet for the research reports and miscellaneous papers generated by the scientists at the laboratory. In 1972, *Publications in Climatology* became a joint monograph of C. W. Thornthwaite Associates Laboratory of Climatology and the University of Delaware Center for Climatic Research under the editorship of Russ Mather, one of the authors of this biography. The monograph series is still being published, with Volume 48 being issued in 1995. Because no research has been conducted at the laboratory since 1970, the series now publishes monograph-length material from outside authors after review and acceptance by an editorial review board.

The Laboratory of Climatology, still located in the building in Centerton, continues with its instrument development work. Now forty-seven years old, the institute for climatic research and training envisioned by Thornthwaite more than fifty years ago still exists, but the changes that have occurred in the academic and research areas have made the need for such an institute less important. Because climatological research and teaching is being conducted at many universities, government agencies, and private research organizations, there is less need for a single center to supply the research and training that Thornthwaite envisioned in 1941.

In 1973, an entire report in *Publications in Climatology* was devoted to a bibliography of the applications of the water budget in physical geography. The four sections—applications to biotic problems, hydrology, soils, and geomorphology—contained references to such articles as "Estimating Yield Components of Wheat from Calculated Soil Moisture" (Baier and Robertson 1967); "The Effect of Water Availability on Tea Yields in Uganda (Hanna 1971); "A Water Yield Model for Small Watersheds" (Haan 1972); "A Climatic Moisture Index for Land and Soil Classification in Canada" (Sly 1970); and "Temperature and Water Content as Factors in Desert Weathering" (Roth 1965). The many references to ongoing research that applied the factors of the water budget to the solution of practical problems strongly suggest that Leighly's negative views about the utility of the water budget approach were incorrect and misinformed.

During the past two decades, with the burgeoning interest in climate change and, especially, in the impact of climate change on water resources, Thornthwaite's water budget has seen ever-increased use. Scientists have used the model in estimating future water supplies to the Great Lakes—for instance, "Impacts of CO_2 Induced Climatic Change on Water Resources in the Great Lakes Basin (Cohen 1986) and "Laurentian Great Lakes Double CO_2 Climate Change Hydrological Impacts" (Croley 1990). Gleick (1986) also used the model in a California study reported in "Methods for Evaluating the Regional Hydrologic Impacts of Global Climatic Changes."

The present authors have undertaken a survey of the *Science Citation Index,* which gives the number of citations to scientists' work. Thornthwaite's citations were found to average fifty-three each year from 1980 to 1984, fifty-eight each year from 1985 to 1989, and, in the last three-year period, from 1990 to 1992, sixty-two each year. In an analysis of citations of English-speaking geographers for the period 1984–1988, Bodman (1991) indicated that Thornthwaite placed fourteenth, just ahead of his mentor and friend, Carl Sauer. In Bodman's list of physical geographers, Thornthwaite ranked sixth.

In the recently published *Encyclopedia of Climatology,* there were twenty-two references to H. H. Lamb and G. T. Trewartha, twenty-one to Thornthwaite, seventeen to R. A. Bryson, and fifteen to J. G. Lockwood. In a survey of climatology texts published over the past three decades, discussions of some aspects of Thornthwaite's work were found in twenty-one of twenty-eight books. An examination of physical geography texts published over the past forty years showed some discussion of Thornthwaite's contributions (usually to the water budget or the moisture index) in thirty-three of fifty-five books. This is truly a remarkable number of references for a geographer who has been dead for more than thirty years and whose work "was seriously questioned in his lifetime and may now be considered obsolete," according to Leighly (1964). Rather, the results show a continuing strong interest in his work and an increasing utilization of his water budget approach.

Rather surprisingly, Thornthwaite has received attention recently for his early interest in urban geography. At the 1995 annual meeting of the Association of American Geographers, geographers James Wheeler and Stanley Brunn presented a paper entitled "Thornthwaite: A Forgotten Pioneer in Urban Geography." In the paper the authors reviewed Thornthwaite's nearly seventy-year-old urban geography dissertation on Louisville, Kentucky, and concluded that many of the approaches he introduced were unique and twenty to thirty years ahead of their time. The authors wondered how differently the field of urban geography in the United States might have developed if Thornthwaite had pursued this early interest in urban geography rather than his beloved climatology. Thornthwaite probably never realized the significance of his contributions to urban geography.

Some colleagues of Thornthwaite's have written to the authors their impressions of Thornthwaite. For example, George Carter of Johns Hopkins University recalled,

> I recall Warren at a meeting of geographers in Washington. He ambled up to the podium and announced in that deep slow voice of his that he had left his speech at home and he guessed that he would just have to talk. He proceeded to give a fascinating talk on microclimatology, with the Seabrook work as his example. I have never been sure whether he had actually written out a speech, knowing that he was fully capable of just talking and making a superior presentation.
>
> On one occasion Warren commented that his wife sent him off with: "Run along now and give your speech," the wifely implication being that he had but one speech. It tickled Warren's sense of humor.
>
> Eventually, when I was head of the department at Johns Hopkins, I added Warren to the faculty. His work in microclimatology and in evapotranspiration was so original and at the same time so practical that it seemed to me to be a wonderful opportunity to combine this with University training of geographers. What a wonderful laboratory, to have Seabrook Farms with its crops and fields to work with.
>
> I recall visiting there several times. The experiment with waste water from the packing plant was a stroke of practical genius, the sort of thing that typified Warren. He suggested spraying the water, thousands of gallons a day on some low sand hills and letting percolation and natural bacterial action purify the water. It seemed like a far-fetched idea but it worked to perfection, clarifying the water, solving a downstream pollution problem, restoring the ground water, and all at a minimum of expense. I think of it as vintage Thornthwaite.
>
> Warren Thornthwaite was an unforgettable man, both as a person and as a scholar with immense creativity combined with an intensely practical bent for solving problems. Such people come at long intervals and are rarely appreciated. Warren was fortunate in that his enormous contributions were recognized and that at Seabrook he had a nearly perfect laboratory to demonstrate the worth of his thinking.

The name C. C. Wallén of Sweden is certainly one well-known in climatological and meteorological circles. Wallén was a friend of Thornthwaite's and wrote of him as follows:

The first time I heard about Warren Thornthwaite must have been in the early 1950s when I learned about his publications from the 1940s on evapotranspiration and his climate classification based upon this concept, in particular the famous one of 1948. In 1954, I spent one year in Mexico as an expert of UNESCO to establish an institute for application of meteorology and hydrology at the University. In this connection, I met several times with Mr. Contreras Arias who was responsible at the Meteorological Service of Mexico to apply Thornthwaite's ideas to the climatology of Mexico. Contreras Arias certainly was a great admirer of Thornthwaite.

My first acquaintance with Warren Thornthwaite came in 1957. In 1951 at the first meteorological Congress in Paris, Thornthwaite had been elected the first President of the Commission for Climatology of the World Meteorological Organization. In fact it was in competition with my own Chief at the time, Dr. A. Angström, Director of the Swedish Meteorological and Hydrological Institute, that he had been elected. I am in fact pretty sure that Dr. Angström was the person who first told me about the outstanding work by Thornthwaite of whom in fact he was an admirer and good friend.

In 1953, Warren Thornthwaite at the second session of the Commission for Climatology in Washington, D.C., was re-elected President for another four years' period. The second meteorological Congress in 1955 considered the need for standardizing national and regional climatological maps and requested the Executive Committee of WMO to study the desirability of starting a project to produce a World Climate Atlas. The Executive Committee in compliance with this request decided at its seventh session in May 1955 to establish a Working Group on Climatological Atlases. Professor Thornthwaite in his capacity of President of the Commission was elected convenor of this group which included in addition the following persons: Dr. A. Angström (Sweden), Prof. K. Hare (Canada), Prof. S. Jackson (South Africa), Dr. N. Rosenan (Israel).

The seventh session of the Executive Committee considered a first report of the Working Group including provisional specifications for a World Climatic Atlas but decided to reconstitute the group to reconsider the specifications in the light of comments from Member countries and the Commission for Climatology.

In order to reconsider the specifications, the Working Group met in January 1957 in New Jersey at Warren Thornthwaite's Laboratory and home. As Dr. Ångström at that time had retired as Director of the Swedish Meteorological Institute and as I myself had been appointed Chief of the Meteorological Bureau of the Institute he asked me to take his position on the working group. So it happened that I, in January 1957, arrived in New Jersey for the meeting of the Group and met Warren Thornthwaite for the first time. I am sure that I do not need to explain my immediate feelings of impact of a very strong personality who dominated his surroundings as well as his colleagues in presiding over the working group. The group met for five days and very rapidly became very good friends. I have over my years in international work met with hundreds of working groups but, among all of them, this one left a special memory. This was partly due to Warren's strong personality but as much to his way of handling the sessions of the group. I remember very little of the formal sessions but much more of the unofficial ones at luncheons, dinners and parties in Warren's home. Indeed, we lived together during the week, Warren's home being the focus of all our activities much more than the meeting room. Professor Kenneth Hare, with whom I have kept in contact over all the years that have passed since the memorable meeting in January 1957, always reminds me of the happy days we spent together on that occasion. Unfortunately, the other members of the group have now passed away as has its convenor.

Let me add that the specifications that were finally drafted at this meeting and later adopted by the Commission for Climatology at its second session and by the Third Congress in 1959 are still standing with only small modifications. Climate maps of mean temperature and mean precipitation have been published for each one of the world's continents on the basis of these specifications.

Thornthwaite left as president of the Commission for Climatology at its session in Washington, D.C., later in January 1957 and thereafter concentrated on his laboratory activities at Seabrook Farms. I think that I met him once more at the Conference on Arid Lands, organized by Unesco in Paris in May 1960.

Ken Hare wrote to the authors some recollections of his association with Thornthwaite:

> Warren was a true genius, with a remarkable intuitive ability to identify how complexes work. But he was impatient of science, and

very suspicious of formal mathematical logic. Much of what is now seen as the core of physical climatology, bioclimatology and ecosystem theory was clear in his mind, but not in his papers. He neither could (because of an unwillingness to submit to drill) nor would write an orthodox defence of his seminal ideas. Yet he was quite anxious for me to write his scientific ideas down, even though he knew that I would choose a more orthodox style of defence.

I last saw him at his home in the late stages of his fatal illness due to spinal cancer which had killed Denzil a year earlier, and he had been convinced that he would go the same way. He did. I went to his room (he was alone in the house). I found him lying in a drugged state on a set of soil heat flux transducers that were enabling him to judge how much heat was passing into his back in the course of pain-easing diathermy. He told me that he had a problem. He had agreed to write a chapter on Agricultural Meteorology in an American Meteorological Society Monograph, whose editor was Paul Waggoner. He had amassed reprints of all his papers, and asked me if I would join him as co-author of a synthesis of his work. Of course I said "yes," and he at once phoned Paul, who agreed.

I spent a few days with him, and at the Lab, where Bill Superior, Russ Mather, June Yoshioka and Gary Yoshioka were all cast down at Warren's impending demise. When I had made sure I had the necessary materials, I said goodbye to him, telling him I would return when I had a draft of the paper. When I was ready, I rang up June Yoshioka, who told me he had just died.

The chapter, called "The Loss of Water to the Air," was entirely my work, as far as penmanship was concerned. But the ideas were wholly his. It gave me the chance to present what I saw [as] an ordered account of his work, expressed with as much rigor as I could muster. It must stand as my personal tribute to him.

These tributes to Thornthwaite are a measure of the esteem that his colleagues felt for him. They respected his scholarship and liked him as a friend. His kindness to his colleagues, as shown in his post-war food parcels to Rudolf Geiger in Germany, mark him as a gentleman. It is his legacy as a scientist, however, that concerns us here.

He changed the course of climatology in twentieth-century

North America by stressing the fact that climatology is a rigorous discipline, defined by physical laws, and not merely the computing of average values of temperature and precipitation. He emphasized that the study of climatology was not limited to the realm of the geographer but belonged also to the botanist, the hydrologist, and the engineer. Following the advice of his friend Rudolf Geiger, he advocated the study of microclimates, or topoclimatology, as he called it. As president of the Association of American Geographers, he challenged geographers to become more scientific and to use mathematical and physical principles in their research. He was the founder of the field of applied climatology, using his knowledge of climate to solve practical problems of irrigation scheduling, crop harvesting and wastewater disposal. His water budget model, based on his novel concept of potential evapotranspiration, has proven to be useful to scores of scientists within and outside the realm of climatology. It has been used by hydrologists throughout the world as a means of quantifying the hydrologic cycle. It has seen increasing use in recent years as part of the burgeoning literature on the impact of possible climate change on water resources.

Warren Thornthwaite was a practical man, and he would be pleased to know that his ideas are helping modern society to solve practical problems. He possessed that spark of genius, that ability to postulate new ideas and to solve complex problems, that is rarely encountered. He clearly left his mark on climatology, on geography, and on the entire realm of science.

Thornthwaite's Published Works

"Base Map of California" (scale 1:3,000,000). Berkeley: University of California Press, 1926.

"Base Map of Eurasia and Africa" (scale 1:40,000,000). Berkeley: University of California Press, 1927.

"The Cylindrical Equal-area Projection for a New Map of Eurasia and Africa." *University of California Publications in Geography* 2, no. 6 (1927): 211–30.

"A Reconstruction of the Natural Vegetation of Ohio" [review]. *Geographical Review* 18 (1928): 502.

"The Polar Front in the Interpretation and Prediction of Oklahoma Weather." *Proceedings of the Oklahoma Academy of Science* 9 (1929): 93–99.

"The Climates of North America According to a New Classification." *Geographical Review* 21 (1931): 633–55.

"A Method of Determining Distance and Height of Objects by Means of a Camera." *Proceedings of the Oklahoma Academy of Science* 11 (1931): 50–51.

"The Great Plains" [review]. *Geographical Review* 22 (1932): 145–47.

"New Light on Climatic Change" (review) *Geographical Review* 22 (1932): 493–95.

"The Quantitative Determination of Climate" (review). *Geographical Review* 22 (1932): 323–25.

"The Climates of the Earth." *Geographical Review* 23 (1933): 433–40.

"Aridity in Australia" (review). *Geographical Review* 24 (1934): 330–32.

"The Climates of Japan" (review). *Geographical Review* 24 (1934): 494–96.

Internal Migration in the United States (with Helen I. Slentz and Carter Goodrich). Philadelphia: University of Pennsylvania Press, 1934.

"Place-Name Study in the United States" (review). *Geographical Review* 24 (1934): 659–60.

"The Great Plains." In *Migration and Economic Opportunity: The Report of the Study of Population Redistribution,* 202–50. Philadelphia: University of Pennsylvania Press, 1936.

"Climatic Studies and Canadian-American Relations." In *Proceedings of the Conference on Canadian-American Affairs,* 60–68. Boston: Ginn and Company, 1937.

"The Hydrologic Cycle Re-examined." *Soil Conservation* 3 (1937): 85–91.

"The Life History of Rainstorms: Progress Report from the Oklahoma Climatic Research Center." *Geographical Review* 27 (1937): 92–111.

"Microclimatic Studies in Oklahoma and Ohio." *Science* 86 (1937): 100–1.

"The Reliability of Rainfall-Intensity-Frequency Determinations." *Transactions of the American Geophysical Union* 18 (1937): 476–84.

"The Research Program of the Section of Climatic and Physiographic Research." *Soil Conservation* 2 (1937): 218–20, 236.

"The Significance of Climatic Studies in Agricultural Research." *Proceedings of the Soil Science Society of America* (1936), 1 (1937): 475–80.

"Climatic Research in the Soil Conservation Service" (with Benjamin Holzman and David I. Blumenstock). *Monthly Weather Review* 66 (1938): 351–68.

"Recent Achievements of the Section of Climatic and Physiographic Research." *Soil Conservation* 3 (1938): 226–29.

"The Determination of Evaporation from Land and Water Surfaces" (with Benjamin Holzman). *Monthly Weather Review* 67 (1939): 4–11.

"The Role of Evaporation in the Hydrologic Cycle." *Transactions of the American Geophysical Union* 20 (1939): 680–86.

"Atmospheric Moisture in Relation to Ecological Problems." *Ecology* 21 (1940): 17–28.

"A Dew-Point Recorder for Measuring Atmospheric Moisture" (with J. C. Owen). *Monthly Weather Review* 68 (1940): 315–18.

"Patterns on Maps and Drawings by the Carbon Transfer Process" (with C. F. Stewart Sharpe). *Science* 91 (1940): 367–68.

"The Relation of Geography to Human Ecology." *Ecological Monographs* 19 (1940): 343–48.

"A Year of Evaporation from a Natural Land Surface" (with Benjamin Holzman). *Transactions of the American Geophysical Union* 21 (1940): 510–11.

Atlas of Climatic Types in the United States 1900–1939. U.S. Department of Agriculture Miscellaneous Publication 421. Washington, D.C.: Government Printing Office, 1941.

"The Chemical Absorption Hygrometer as a Meteorological Instrument." *Transactions of the American Geophysical Union* 22 (1941): 429–32.

"Climate and Settlement in the Great Plains." In *Climate and Man*, 177–87. (U.S. Department of Agriculture 1941 Yearbook). Washington, D.C.: Government Printing Office, 1941.

"Climate of the Southwest in Relation to Accelerated Erosion" (with C. F. Stewart Sharpe and Earl F. Dosch) *Soil Conservation* 6 (1941): 298–302, 304.

"Climate and the World Pattern" (with David I. Blumenstock). In *Climate and Man*, 97–127. Washington, D.C.: Government Printing Office, 1941.

"Evaporation and Transpiration" (with Benjamin Holzman). In *Climate and Man*, 545–50. Washington, D.C.: Government Printing Office, 1941.

Climate and Accelerated Erosion in the Arid and Semiarid Southwest, With Special Reference to the Polacca Wash Drainage Basin (with C. F. Stewart Sharpe and Earl F. Dosch). U.S. Department of Agriculture Technical Bulletin 808. Washington, D.C.: Government Printing Office, 1942.

Measurement of Evaporation from Land and Water Surfaces (with Benjamin Holzman). U.S. Department of Agriculture Technical Bulletin 917. Washington, D.C.: Government Printing Office, 1942.

"Note on the Variation of Wind with Height in the Layer near the Ground" (with Maurice Halstead). *Transactions of the American Geophysical Union* 23 (1942): 249–55.

"Atmospheric Turbulence and the Measurement of Evaporation." In *Proceedings of the Second Hydraulic Conference* (June 1–4, 1942, College of Engineering, State University of Iowa). Studies in Engineering, Bulletin 27. 280–88. Ames, Iowa, 1943.

"Estado y programa de la climatologia en los Estados Unidos de America." *Boletin de la Sociedad Mexicana de Geografia y Estadistica* 58 (1943): 425–50.

"Meteorology and Climatology" (review article). *Science* 97 (1943): 580–83.

"Problems in the Classification of Climates.' *Geographical Review* 33 (1943): 233–55.

"Status and Prospects of Climatology" (with John Leighly). *Scientific Monthly* 57 (1943): 457–65.

"Wind Gradient Observations" (with Paul Kaser). *Transactions of the American Geophysical Union* 24 (1943): 166–82.

The Economic Relationship of Climate and Agriculture. Comité Permanenta de la Segunda Conferencia Interamericana de Agricultura, México D.F. Publicaciones, sección 7a (1944), 15pp.

"Estúdios acerca de la medida de la evaporación." *Revista de la Sociedad de Estúdios Astronómicos y Geofísicos* (Mexico), 4, no. 3 (1944): 29–35.

"Report of the Committee on Transpiration and Evaporation, 1943–44" (with H. G. Wilm). *Transactions of the American Geophysical Union* 25 (1944): 686–93.

Discussion of "On Evaporation from a Free Water Surface," by G. H. Hickox. *Proceedings of the American Society of Civil Engineers* 71 (1945): 343–55.

Discussion of "On Investigation of Soil Water by Means of Weighing Lysimeters" L. L. Harrold and F. R. Dreibilbis. *Transactions of the American Geophysical Union* 26 (1945): 292–97.

"Report of the Committee on Evaporation and Transpiration, 1944–1945" (with H. G. Wilm). *Transactions of the American Geophysical Union* 26 (1945): 292–97.

"Meteorology" [review]. *Science* 102 (1945): 432–34.

"El Agua en la Agricultura." *Irrigación en México* 27, no. 2 (1946): 19–39.

"The Moisture Factor in Climate." *Transactions of the American Geophysical Union* 27 (1946): 41–48.

"Report of the Committee on Evaporation and Transpiration, 1945–46" (with H. G. Wilm). *Transactions of the American Geophysical Union* 27 (1946): 721–23.

"Air." In *World Book Encyclopedia,* vol. A, 105–9. 1947.

"Climate." In *World Book Encyclopedia,* vol. C, 1486–90. 1947.

"Climate and Moisture Conservation." *Annals of the Association of American Geographers* 37 (1947): 87–100.

"An Approach toward a Rational Classification of Climate." *Geographical Review* 38 (1948): 55–94.

"Climate and Soil Moisture in the Tropics." *Geographical Review* 39 (1949): 498–501.

"Evaporation and Transpiration" (with Adolph F. Meyer). In *Hydrology Handbook,* 119–40. American Society of Civil Engineers, Manual of Engineering Practice, no. 28. 1949.

"Recent Climatic Studies in Australia." *Geographical Review* 39 (1949): 671–73.

"Report of the Committee on Climatology, 1947–48" (with E. R. Biel, P. E. Church, W. C. Jacobs, H. Landsberg, J. B. Leighly, and K. Hafsted). *Transactions of the American Geophysical Union* 30 (1949): 440–42, 445.

"Climate." In *Collier's Encyclopedia* 13:468–74. 1950.

"Modern Meteorology and Its Application to the Agricultural Regions of the United States." In *Water and Man,* edited by Jonathan Forman and Ollie E. Fink, 74–83. Columbus, Ohio: Friends of the Land, 1950.

"Agricultural Climatology at Seabrook Farms." *Weatherwise* 4 (1951): 27–30.

"Meteorology." In *World Book Encyclopedia 1951 Annual Supplement*, 154. 1951.

"Report of the Committee on Climatology, 1950–1951" (with Phil E. Church and John R. Mather). *Transactions of the American Geophysical Union* 32 (1951): 765–68.

"The Role of Evapotranspiration in Climate" (with John R. Mather). *Archiv für Meteorologie, Geophysik und Bioklimatologie* ser. B, 3 (1951): 16–39.

"Rudolf Geiger and the Science of Microclimatology" (review). *Geographical Review* 41 (1951): 162–64.

"The Water Balance in Tropical Climates." *Bulletin of the American Meteorological Society* 32 (1951): 166–73.

"A Charter for Climatology." *World Meteorological Organization* 2 (1952): 40–46.

"Climate in Relation to Planting and Irrigation of Vegetable Crops." In *Proceedings of the Eighth General Assembly and Seventeenth International Congress, Washington, 1952, International Geographical Union,* 290–95. Washington, D.C.: U.S. National Committee of the International Geographical Union. Also *Publications in Climatology* 5, no. 5 (1952).

"Climate and Scientific Irrigation in New Jersey." *Publications in Climatology* 6 (1952): 1–8.

"Crops by Slide Rule." In *The Story of Our Time, Encyclopedia Year Book, 1953,* 290–92. 1952.

"Drought." In *Encyclopædia Britannica* 7, 674. 1952.

"Evapotranspiration in the Hydrologic Cycle." In *The Physical Basis of Water Supply and Its Principal Uses, the Physical and Economic*

Foundation of Natural Resources, 25–35. Interior and Insular Affairs Committee, House of Representatives, U.S. Congress. Washington, D.C.: Government Printing Office, 1952.

"Grassland Climates." *Publications in Climatology* 5, no. 6 (1952): 1–14.

"Klima-Kalender und Pflanzenwachstum." *Die Umschau* 53 (1952): 522–24.

"Meteorology." In *World Book Encyclopedia 1952 Annual Supplement,* 122. 1952.

"Operations Research in Agriculture." *Journal of the Operations Research Society of America* 1 (1952): 33–38.

"Climate and Scientific Irrigation in New Jersey." *Publications in Climatology* 6 (1953): 1–8.

"Meteorology." In *World Book Encyclopedia 1953 Annual Supplement,* 158. 1953.

"The Place of Supplemental Irrigation in Post-War Planning." *Publications in Climatology* 6 (1953): 9–29.

"The Water Balance in Arid and Semiarid Climates." In *Desert Research* (Proceedings of the International Symposium held in Jerusalem, May 7–14, 1952), 112–35. Research Council of Israel, Special Publication 2. Jerusalem, 1953.

"Climate in Relation to Crops." (with John R. Mather). In "Recent Studies in Bioclimatology: A Group." *Meteorological Monographs* 2, no. 8 (1954): 1–10.

"The Development of Anemometers for Observing Wind Gradients." In "Final Report, Micrometeorology of the Surface Layer of the Atmosphere: The Flux of Momentum, Heat and Water Vapor." *Publications in Climatology* 7 (1954): 271–276.

"Grow Plants on a Timetable." *Organic Gardening and Farming* 1 (1954): 53–55.

"Guide to Climatic Classification" [review]. *Geographical Review* 44 (1954): 437–38.

"I Believe . . . that We Should Overhaul Our Ideas on What Makes a 'Good' Weather-Forecasting Service for Farmers." *Country Gentleman* 124 (1954): 11–12.

Introduction to "Final Report, Micrometeorology of the Surface Layer of the Atmosphere: The Flux of Momentum, Heat, and Water Vapor." *Publications in Climatology* 7 (1954): 235–49.

"The Measurement of Humidity Profiles" (with Owen Beenhouwer). *Publications in Climatology* 7: 261–70.

"A Re-examination of the Concept and Measurement of Potential Evapotranspiration." In "The Measurement of Potential Evapotranspiration," edited by John R. Mather. *Publications in Climatology* 7 (1954): 200–9.

Review of *Climate, Vegetation and Man,* by Leonard Hadlow. *Scientific Monthly* 78 (1954): 51–52.

Review of *An Introduction to Climate,* by G. T. Trewartha. *Science* 120 (1954): 1067–68.

"Topoclimatology." In *Proceedings of the Toronto Meteorological Conference 1953* 227–32. London: Royal Meteorological Society, 1954.

"We Must Irrigate Scientifically." *American Vegetable Grower* 2, nc. 6 (1954): 51–52.

"Climatic Classification in Forestry" (with F. Kenneth Hare). *Unasylva* 9 (1955): 51–59.

"Climatology and Irrigation Scheduling" (with John R. Mather). *Weekly Weather and Crop Bulletin* 42, no. 26 (1955): 6–8.

"Discussions on the Relationship between Meteorology and Oceanography (with G. E. R. Deacon, H. V. Sverdrup, H. Stommel). An Oceanographic Convocation Held at Woods Hole, Mass., June 22, 23, 24, 1954." *Sears Foundation Journal of Marine Research* 14 (1955): 510–15.

Review of "Our American Weather," by G. H. T. Kimble. *Scientific Monthly* 80 (1955): 386.

Review of "Tropical Soils," by E. C. J. Mohr and F. A. van Baren. *Science* 121 (1955): 670.

"The Water Balance" (with John R. Mather). *Publications in Climatology* 8 (1955): 1–104.

"The Water Budget and Its Use in Irrigation" (with John R. Mather). In *Water: The Yearbook of Agriculture 1955,* 346–58. Washington, D.C.: U.S. Department of Agriculture, 1955.

"The Air as a Water Absorbing Medium." In *Handbuch der Pflanzenphysiologie* (Encyclopedia of Plant Physiology), edited by W. Ruhland, 3: 257–64. Berlin: Springer Verlag, 1956.

"Climatology in Arid Zone Research." In *The Future of Arid Lands,* edited by Gilbert F. White, 67–84. American Association for the Advancement of Science Publication 43. Washington, D.C.: AAAS, 1956.

"Microclimatic Investigations at Point Barrow, Alaska" (with John R. Mather) *Publications in Climatology* 9 (1956): 1–51.

"Modification of Rural Microclimates." In *Man's Role in Changing the Face of the Earth,* edited by William L. Thomas, 567–83. Chicago: University of Chicago Press, 1956.

"Dew Point Apparatus" (with H. Hacia). In *Exploring the Atmosphere's First Mile,* vol. 1, *Instrumentation and Data Evaluation,* edited by Heinz H. Lettau and Ben Davidson, 176–82. New York: Pergamon Press, 1957.

"Instructions and Tables for Computing Potential Evapotranspiration and the Water Balance" (with John R. Mather and D. B. Carter). *Publications in Climatology* 10 (1957): 181–311.

"Modified SCS Cup Anemometers" (with M. H. Halstead). In *Exploring the Atmosphere's First Mile,* vol. 1, *Instrumentation and Data Evaluation,* edited by H. H. Lettau and B. Davidson, 128–33. New York: Pergamon Press, 1957.

"The Task Ahead in Climatology." *World Meteorological Organization Bulletin* 6 (1957): 2–7.

"Estimating Soil Moisture and Tractionability Conditions for Strategic Planning" (with C. E. Molineux, John R. Mather, and D. B. Carter). *Air Force Surveys in Geophysics,* 94 (1958).

"Introduction to Arid Zone Climatology." In *Climatology and Microclimatology: Proceedings of the Canberra Symposium* (UNESCO, *Arid Zone Research,* vol. 11), 15–22. 1958.

"Microclimatic Investigations at Point Barrow, Alaska, 1956–1958" (with John R. Mather). *Publications in Climatology* 11 (1958): 59–239.

"Three Water Balance Maps of Southwest Asia" (with John R. Mather and D. B. Carter). *Publications in Climatology* 11 (1958): 1–57.

"Water Investigations in New Jersey." In *Water Supply: Papers Delivered at a Meeting of the Princeton University Conference, Jan. 29–30, 1958,* 85–97. Princeton, N.J.: Princeton University Conference, 1958.

"Investigations of the Climatic and Hydrologic Factors Affecting the Redistribution of Strontium-90 in the Soil" (with John R. Mather). *Publications in Climatology* 12 (1959): 49–91.

"Measurement of Vertical Winds in Typical Terrain" (with W. J. Superior, F. K. Hare, and K. R. Ono). *Publications in Climatology* 12 (1959): 93–204.

"Equation and Table for Determination of the Wave of Leaching in the Soil" (with Sally Thornthwaite). *Publications in Climatology* 13 (1960): 155–58.

"Movement of Radiostrontium in Soils" (with John R. Mather and J. K. Nakamura). *Science* 131 (1960): 1015–19.

"American Geographers: A Critical Evaluation." *The Professional Geographer* 13 (1961): 10–12.

"The Measurement of Vertical Winds and Momentum Flux" (with W. J. Superior, John R. Mather, and F. K. Hare). *Publications in Climatology* 14 (1961): 1–89.

"The Task Ahead." *Annals of the Association of American Geographers* 51 (1961): 345–56.

"Vertical Winds near the Ground at Centerton, N.J." (with W. J. Superior and John R. Mather). *Publications in Climatology* 14 (1961): 91–244.

"The Geographer's Role in Climatology." In *Hermann von Wissmann—Festschrift,* edited by Adolf Leidlmair, 81–88. Tübingen, Germany: Geographisches Institut der Universität Tübingen, 1962.

Bibliography

Baier, W. and G. W. Robertson. "Estimating Yield Components of Wheat from Calculated Soil Moisture." *Canadian Journal of Plant Science* 47, no. 7 (1967): 617–30.

Black, P. E. "Streamflow Increases following Farm Abandonment in Eastern New York Watershed." *Water Resources Research* 4, no. 6 (1968): 1171–77.

Blumenstock, D. *The Ocean of Air*. New Brunswick, N.J.: Rutgers University Press, 1959.

Bodman, A. R. "Weavers of Influence: The Structure of Contemporary Geographic Research." *Transactions of the Institute of British Geographers*, n.s., 16 (1991): 21–37.

Carter, D. B., D. W. Bridge, R. R. Clark, D. O. Inboden, J. J. Swartz, and T. P. Tripp. "Application of the Water Budget in Physical Geography, An Annotated Bibliography." *Publications in Climatology* 26, no. 3 (1973): 1–26.

Cohen, S. J. "Impacts of CO_2 Induced Climatic Change on Water Resources in the Great Lakes Basin." *Climatic Change* 8, no. 2 (1986): 135–54.

Croley, T. E., II "Laurentian Great Lakes Double CO_2 Climate Change Hydrological Impacts." *Climatic Change* 17 (1990): 27–47.

Garnier, B. "The Climates of New Zealand According to Thornthwaite's Classification." *Annals of the Association of American Geographers* 36, no. 3 (1946): 151–77.

Geiger, R. *The Climate Near the Ground*. Translated by Milroy Stewart. Cambridge, Mass.: Harvard University Press, 1950.

Geiger, R., and H. Zierl. "Köppens Klimazonen und die Vegetationszonen von Africa." *Gerlands Beiträge zur Geophysik* 33 (1931): 292–304.

Gleick, P. H. "Methods for Evaluating the Regional Hydrologic Impacts of Global Climatic Changes." *Journal of Hydrology* 88 (1986): 97–116.

Haan, C. T. "A Water Yield Model for Small Watersheds." *Water Resources Research* 8, no. 1 (1972): 58–69.

Hanna, L. W. "The Effect of Water Availability on Tea Yields in Uganda." *Journal of Applied Ecology* 8 (1971): 791–813.

Hare, F. K. "Charles Warren Thornthwaite, 1899–1963." *Geographical Review* 53, no. 4 (1963): 595–97.

Higgins, J. J. "Instructions for Making Phenological Observations of Garden Peas." *Publications in Climatology* 5, no. 2 (1952): 1–11.

Holzman, B. *Sources of Moisture for Precipitation in the United States.* USDA Technical Bulletin 589. Washington, D.C.: Government Printing Office, 1937.

Köppen, W. "Versuch einer Klassifikation der Klimate, vorzugsweise nach ihren Beziehungen zur Pflanzenwelt." *Geograph. Zeitschrift* 6 (1900): 593–611 and 657–79.

Köppen, W. *Die Klimate der Erde.* Berlin: Walter De Gruyter and Company, 1923.

Köppen, W. "Das Geographische System der Klimate." *Handbuch der Klimatologie* 1 (1936): 1–44.

Leighly, J. "Charles Warren Thornthwaite, March 7, 1899–June 11, 1963." *Annals of the Association of American Geographers* 54 (1964): 615–21.

Leighly, J. "Carl Ortwin Sauer, 1889–1975." *Annals of the Association of American Geographers* 66, no. 3 (1976): 337–46.

Leighly, J. "Drifting into Geography in the Twenties." *Annals of the Association of American Geographers* 69, no. 1 (1979): 4–15.

Marcus, M. G. "Coming Full Circle: Physical Geography in the Twentieth Century." *Annals of the Association of American Geographers* 69 (1979): 521–32.

Marsh, G. P. *Man and Nature; or, Physical Geography as Modified by Human Action.* New York: Charles Scribner, 1864.

Mather, J. R. "The Disposal of Industrial Effluent by Woods Irrigation." *Transactions of the American Geophysical Union* 34, no. 2 (1953): 227–39.

Mather, J. R., ed. "Thornthwaite Memorial Volume I, Papers on Evapotranspiration and the Climatic Water Balance; II, Papers on Selected Topics in Climatology." *Publications in Climatology,* 25, nos. 2 and 3 (1972).

McDougal, E. "The Moisture Belts of North America." *Ecology* 6 (1925): 325–32.
Oliver, J. E. and R. W. Fairbridge. *The Encyclopedia of Climatology, Vol. XI: Encyclopedia of Earth Sciences*. New York: Van Nostrand Reinhold Company, 1987.
Réaumur, R. A. F. de. "Observations du thermomètre, faites à Paris pendant l'année 1735, comparées avec celles qui ont été faites sous la ligne, à l'Isle de France, à Alger et en quelques-unes de nous isles de l'Amérique." *Paris Mem., Acad. Sci.*, 1735.
Roth, E. W. "Temperature and Water Content as Factors in Desert Weathering." *Journal of Geology* 73 (1965): 454–68.
Russell, R. J. "Climates of California." *Publications in Geography* 2, no. 4 (1926): 73–84.
Russell, R. J. "Dry Climates of the United States: I. Climate Map." *Publications in Geography* (University of California), 5, no. 1.
Sanderson, M. "Climate Change and Water in the Great Lakes Basin." *Canadian Water Resources Journal* 18 (1993): 417–24.
Sauer, C. O. "The Morphology of Landscape." *Publications in Geography* (University of California), 2, no. 2 (1925): 19–53.
Seabrook, J. M. "Applied Climatology at Seabrook Farms." *Weatherwise*, April 1953, 36–37.
Slentz, Floyd. Central Michigan University, Development Office. Typed notes for Alumni Banquet, October 15, 1988.
Sly, W. K. "A Climatic Moisture Index for Land and Soil Classification in Canada." *Canadian Journal of Soil Science* 50 (1970): 291–310.
Stanislauski, D. "Carl Ortwin Sauer, 1889–1975." *Journal of Geography*, December 1975, 548–54.
Thomas, W. L., ed. *Man's Role in Changing the Face of the Earth*. Chicago: University of Chicago Press, 1956.
Van Royen, W. "The Climatic Regions of Eastern North America." *Monthly Weather Review* 55 (1927): 410–12.

Index

Abbé, Cleveland, 53
Advisory Committee on Arid Zone Research, UNESCO, 161
Air Weather Service (AWS), U.S. Air Force, 77, 78, 79, 141, 143
Alagoz, Halis, 178
Albani, F., 173
Ali, Farag Mohamed, 174
American Geographical Society, 151, 180, 181 188
Angström, Anders, 129, 161, 173, 175, 199, 200
Annals of the Association of American Geographers, 47, 59, 67, 68, 190
Ann Arbor, Mich., 46, 47
Arakelian, Lt. Col., Air Weather Service, 143
Arctic Institute of North America, 99, 136
Argentina, 104, 155, 173
Arkansas, state of, 132, 133, 134
Arlington, Va., 37, 50, 135
Ashford, Oliver, 177, 178
Association of American Geographers, 13, 39, 46, 47, 59, 67, 68, 151, 172, 182, 184, 190, 197, 202

Bad Kissingen, Germany, 61, 178
Baier, Wolfgang, 195
Baker, O. E., 35, 140
Bay City, Mich., 3, 4, 5, 11
Bennett, H. H., 36, 41, 42
Berkner, Lloyd, 181
Biel, Erwin, 155
Blumenstock, David 44, 45, 48, 49, 50, 57, 58
Bollinger, C. J., 17, 18, 20, 23, 24, 34, 37

Bowman, Isaiah, 15, 35, 36, 39, 40, 41, 46, 59, 78, 79, 146
Bowman School of Geography, Johns Hopkins University, 141–47
Bronk, Detlev W., 145
Brookings Institute, Washington, D.C., 36
Brooks, C. E. P., 173
Bryson, R. A., 197
Brunn, Stanley, 197
Budyko, M. I., 193
Burrill, Peter, 52

C. W. Thornthwaite Associates Laboratory of Climatology, 53, 95, 100, 102, 104, 111, 125–28, 131, 132, 135, 139, 141, 142, 143, 145–50, 156, 167, 181, 188, 192–95. *See also* Drexel Institute of Technology
Calkins, R. D., 10, 11, 48
Campbell Soup Company, 123
Candolle, A. de, 26
Carey, Joe, 10
Carter, Doug, 82, 142, 143, 185
Carter, George, 15, 51, 78, 144, 145, 198
Carter, Murphy, 185
Centerton, N.J., 100, 127, 148, 149, 150, 161, 168, 169, 174, 178, 190, 192, 195
Central Michigan Normal School, 7, 8, 9, 11, 185, 186, 187, 189
Charles Warren and Denzil Slentz Thornthwaite Memorial Scholarship, 185, 186, 187
Chesapeake Bay Institute, 145
Church, Phil, 155
Clare County, Mich., 7, 8, 9, 57
Climatic and Physiographic Division of Soil Conservation Service, U.S. Department of Agriculture, 36, 37, 39, 47, 55, 75, 107, 140, 180, 188
Cohen, S. J., 196
College Park, Md., 57, 140
Collins, Henry, 39
Commission for Climatology (CCL), WMO, 157–60, 163, 165, 174, 188, 200
Commonwealth Scientific and Industrial Organization CSIRO, Australia, 168, 191
Contreras, Arias A., 173, 178, 199

Country Gentleman magazine, 127
Court, Arnold, 20, 22
Covey, Winton, 82
Croley, T., 196
Cullman Brothers, 127, 128, 130, 131
Cullum Geographical Medal, 180, 181, 188
Curry, Leslie, 175

Darwin, Charles, 25
Davies, D. A., 177
Davis, William Morris, 12, 13
Deij, L. J. L., 173
Del Monte Packing Corp., 123
Dicken, Sam, 13, 16
Dodge, Stanley, 13, 48
Dordick, Isadore, 32
Dove, H. W., 25
Drexel Institute of Technology, Laboratory of Climatology, 100, 125, 149, 150, 168, 174, 181, 188, 194 *See also* C. W. Thornthwaite Associates Laboratory of Climatology
Dzerzeevski, Dr. (Russian climatologist), 191

East Lansing, Mich., 182, 184
Eberswalde, Germany, 61
Effluent Disposal Branch, C. W. Thornthwaite Associates, 127
Erlangen, Germany, 60, 61
Essex (later Hudson), Fidelia, 4
Essexville, Mich., 3, 4, 5
Esteros Point, Calif., 30

Fairchild, Wilma, 153, 154
Flohn, H., 173, 178
Food and Agricultural Organization (FAO), United Nations, 156, 159, 160
Foust, Judson W., 186, 189
Frankfurt, Ky., 15, 23
Frost, Ruel, 17
Fukuda, H., 173

Garnier, Ben, 168
Geiger, Hans, 60
Geiger, Rudolf, 24, 33, 60, 61, 142, 167, 169, 170, 191, 192, 201, 202
Gentilli, J., 172
Gleick, P. H., 196
Godske, C. L., 178
Grand Rapids, Mich., 59
Graz, Austria, 25, 29
Grisebach, A., 25
Grossborstel, Germany, 25
Guerrini, V. H., 177

H. J. Heinz Company, 123, 172
Haan, C. T., 196
Hafstad, Katherine, 155
Halstead, Maury, 51, 81, 82, 112, 142, 144, 175
Hamburg, Germany, 24, 25
Hamid, Q., 174
Hammonton, N.J., 126
Hanna, L. W., 196
Hare, Kenneth, 65, 159, 161, 176, 190, 199, 200
Hartshorne, Richard, 48, 82
Hawaii, state of, 44
Hewes, Leslie, 20, 22
Higgins, J. J., 94, 95
Hitchcock, Charles, 181
Hla, Maung, 168
Hobbs, W. H., 10
Holzman, Benjamin, 47, 48, 50, 63; in Air Weather Service, 77, 79, 155
Hong Kong, 104
Hudson, Henry, 3
Hudson, Joseph, 4, 5
Hyattsville, Md., 57

Ickes, Harold, 36
International Geophysical Union (IGU), 158, 159
International Geophysical Year, 136
International Meteorological Organization (IMO), 156, 157, 167

Isaiah Bowman School of Geography, Johns Hopkins University, 15, 141, 143–47
Israel, 104, 155, 161, 173, 174, 177, 199

Jackson, A. V. Williams, 60
Jackson, Ian 178
Jackson, Justin, 137
Jackson, Stanley, 161, 177, 199
Jacobs, W. C., 155, 174
James, Preston, 13, 48
Jefferson, Mark, 10
Jobbeous, George, 168
Johns Hopkins Laboratory of Climatology, 125, 127, 147, 194. *See also* C. W. Thornthwaite Associates Laboratory of Climatology
Johns Hopkins University, 15, 36, 51, 59, 78, 79, 82, 125, 127, 141, 143, 147, 149, 168, 170, 181, 188, 198. *See also names of schools, departments, etc.*
Johnson, Francis, 40, 41
Jones, Wellington, 13

Kamph, A. H., 173
Kapuskasing, Ont., 155, 156
Karachi, Pakistan, 162, 166
Keil, Karl, 178
Kentucky Geological Survey, 15, 23
Khan, Barket Ullah, 177
Klippel (later Geiger), Irmgard, 60
Kniffen, Fred, 13, 55
Köppen, Marie, 25
Köppen, Wladimir, 12, 16, 20, 24–30, 32, 33, 58, 60
Kraus, E., 177

Laboratory of Climatology, Seabrook, N.J., 53, 82, 95, 99, 100, 102, 104, 111, 125, 135, 141, 142, 143, 145–50, 156, 159, 167, 181, 188, 193, 194, 195. *See also* Drexel Institute of Technology, Laboratory of Climatology; C. W. Thornthwaite Associates Laboratory of Climatology
Lamb, H. H., 197
Landsberg, Helmut, 155, 174

Land Use Committee, Science Advisory Board, U.S. Department of the Interior, 36, 39
Laycock, Arleigh, 176
Lee, Douglas H. K., 170
Leighly, John, 9–13, 15, 30, 39, 44, 47, 48, 51, 52, 53, 57, 58, 59, 72, 74, 75, 82, 155, 168, 170, 183, 190, 196, 197
Leipzig University, 24, 25
Lettau, Heinz, 170
Lockwood, J. G., 197
Louisiana State University, 37
Louisville, Ky., 15, 16, 135, 197
Lowdermilk, W., 36, 38, 41
Lustig (now Sanderson), Marie, 103, 140, 156

Manley, Gordon, 174
Marcus, Melvin, 184
Marsh, George Perkins, 12, 175
Martin, Geoffrey, 39
Mather, John R. (Russ), 81, 82, 97, 104, 126, 127, 142, 144, 150, 176, 194, 195, 201
McDougall, Eric, 28
McGill University, 65, 168, 178
McMurry, Kenneth, 13, 48
Meigs, Peveril, 13, 14, 159
Metcalf, Ebeneezer, 4
Mexico, 62, 75, 76, 100, 104, 156, 173, 199
Mexico City, 155, 178
Michigan Land Economic Survey, 10
Miller, A. A., 20, 173
Miller, David, 62
Moller, W., 162
Monticello, Ohio, 8
Mount Pleasant, Mich., 7–11, 187, 189
Mrs. Paul's (food processing company), 123
Muskegon, Mich., 57
Muskingum, Ohio, 43, 44, 48, 50

Nebiker, Walter, 178
Nigeria, 104

Norman, Okla., 17, 19, 20, 38
Normand, Dick, 40
Northwest Territories, Canada, 156

Office of Naval Research, U.S., 99, 171, 172, 175, 176, 178, 182
Oklahoma Academy of Science, 23, 188
O'Neill, Nebr., 102, 174
Ono, K. Ray, 137, 138
Ontario Research Foundation, 141
Orvig, Svenn, 178
Owosso, Mich., 11

Page, John L., 17
Parmelee, D. W., 127, 194, 195
Pearce, Webster, 11
Penck, Albert, 13, 31
Penman, Howard, 65, 174
Perkins, J. A., 149
Petterssen, Sverre, 192
Phillips, John, 178
Pinconning, Mich., 4, 6
Point Barrow, Alaska, 176, 179
Polacca Wash Project, Ariz., 39, 40, 41
Portman, Donald, 82, 142
Powers, Harold, 11
Price, Saul, 44
Priestley, C. H. B., 168

Ramdas, L. A., 191
Réaumur, René, 84, 85, 87
Reiche, Perry, 40, 41
Robertson, George, 196
Rockefeller, W. A., 181
Roop, J. C., 181
Roosevelt, Franklin D., 36
Rosenan, Naftali, 161, 173, 177, 199
Rothamsted, England, 65, 174
Russell, R. J., 28, 30, 31, 33, 37, 49

Salisbury, R. D., 10
Sauer, Carl, 10–16, 20, 22, 23, 24, 32, 36, 39, 40, 41, 43, 44, 45, 48, 49, 50, 57, 59, 197
Scientific Monthly, 53, 59, 77
Seabrook, C. F., 76, 81, 106
Seabrook, Jack, 75, 77, 80, 81, 106, 108, 109, 118
Seabrook, N.J., 66, 72, 83, 90, 101, 109, 141, 142, 144–49, 155, 193, 198
Seabrook Farms Company, 75–79, 81–86, 88, 90, 93, 97, 99, 100, 101–8, 111, 112, 114, 115, 116, 119, 122, 123, 125, 127, 131, 141, 142, 144, 145, 171, 172, 198, 200
Sears, Paul, 24, 38
Sekiguti, T., 172, 173
Semple, Ellen Churchill, 12
Shantz, Homer, 66–70, 102
Sharpe, C. F. Stewart, 47
Siple, Paul, 62
Slatyer, H. O., 177
Slentz, Eunice Ann, 8, 9, 14, 15
Slentz, Floyd, 4, 8, 92, 125, 126, 139, 194
Slentz, Helen Irene, 8, 19
Sly, W. K., 196
Soil Conservation Service, U.S. Department of Agriculture, 34, 36, 37, 38, 41, 44, 46, 47, 50, 55, 57, 59, 60, 75, 76, 77, 107, 108, 140, 180, 188. *See also names of individual employees*
Subrahmanyam, V. P., 173
Superior, W. J., 137, 138, 194, 201
Swarbrick, James, 161, 162, 178
Swoboda, G., 160

Taylor, Griffith, 13
Tehran, Iran, 162, 179
Ten Kate, H., 173
Thaller, M. X., 174
Thornthwaite, Calvin, 3
Thornthwaite, Charles Warren, 3,
Thornthwaite, Denzil (née Slentz), 7, 8, 9, 14–19, 22, 57, 165, 184, 185, 186, 201
Thornthwaite, Dorothy, 37, 186
Thornthwaite, Edward, 3

Thornthwaite, Elizabeth (Higgins-Hallway), 12, 14, 17, 19, 22, 57, 86, 186, 187
Thornthwaite, England, 3
Thornthwaite, Ernest, 3, 4, 7
Thornthwaite, Faith, 5, 11
Thornthwaite, Fred, 5, 16
Thornthwaite, Harry, 3
Thornthwaite, Isabelle, 3
Thornthwaite, Mildred (née Hudson) (Thornthwaite's mother), 3, 4, 5, 7
Thornthwaite, Mildred (Thornthwaite's sister), 3, 4, 5
Thornthwaite, Sally, 37, 186
Thornthwaite, William, 3
Toronto, Canada, 103, 141, 155, 156, 175
Trewartha, Glenn, 197
Troll, K., 173

United Nations Educational, Scientific, and Cultural Organization (UNESCO), 156, 159, 160–63, 165, 166, 178, 179, 188, 191, 192, 199, 200
United States Army: Corps of Engineers, 99; Signal Corps, 99, 172; Quartermaster Corps, 62, 77, 136
United States Weather Bureau, 28, 37, 38, 44, 53, 54, 73, 82, 107, 161
University of Arkansas, 133
University of California–Berkeley, 10, 11, 16, 23, 44, 187
University of Chicago, 10, 44, 150, 187, 188
University of Delaware, Center for Climatic Research, 148, 149, 194
University of Maryland, 77, 140, 188
University of Michigan, 10, 11, 13
University of Oklahoma, 17–20, 23, 24, 34, 38, 140, 151
University of Pennsylvania, Study of Population Distribution, 22, 33, 34, 35
University of Toronto, 13
University of Wisconsin, 62

Van Royen, W., 28
Von Ficker, H., 157
Von Humboldt, Alexander, 25, 58
Von Wissmann, H., 172, 177

Waggoner, Paul, 201
Wallén, C. C., 174, 198

Ward, R. de C., 30
Washington, D.C., 34, 36, 44, 45, 49, 52, 55, 56, 59, 62, 76, 107, 108, 127, 140, 157, 163, 198, 199, 200
Washington Bookshop Association, 55, 56
Wegener, Alfred, 25
Wheeler, James, 197
Wilcock, A. A., 173
Wood, Walter, 180, 181
Wooldridge, S. W., 173
Works Progress Administration (WPA), 38, 39
World Health Organization (WHO), 156, 157, 159
World Meteorological Organization (WMO), 156, 158, 164, 177, 188, 199. *See also names of agencies, commissions, etc.*
Worrilow, George, 149

Yao, A. Y. M., 177
Yoshioka, Gary, 201
Yoshioka, June, 149, 194, 195, 201
Ypsilanti, Mich., 10, 39